OCCUPATIONAL EXPOSURE TO SILICA

AND CANCER RISK

INTERNATIONAL AGENCY FOR RESEARCH ON CANCER

The International Agency for Research on Cancer (IARC) was established in 1965 by the World Health Assembly, as an independently financed organization within the framework of the World Health Organization. The headquarters of the Agency are at Lyon, France.

The Agency conducts a programme of research concentrating particularly on the epidemiology of cancer and the study of potential carcinogens in the human environment. Its field studies are supplemented by biological and chemical research carried out in the Agency's laboratories in Lyon and, through collaborative research agreements, in national research institutions in many countries. The Agency also conducts a programme for the education and training of personnel for cancer research.

The publications of the Agency are intended to contribute to the dissemination of authoritative information on different aspects of cancer research. A complete list is printed at the back of the book.

Cover illustration: Scanning electron microscope view of microcrystals of ultra-pure quartz prepared by hydrothermal synthesis (by courtesty of 'Centre de Recherche sur la Synthèse et la Chimie des Minéraux'); inset, chest X-ray of subject suffering from silicosis.

WORLD HEALTH ORGANIZATION

INTERNATIONAL AGENCY FOR RESEARCH ON CANCER

OCCUPATIONAL EXPOSURE TO SILICA AND CANCER RISK

Edited by

L. Simonato, A.C. Fletcher, R. Saracci and T. L. Thomas

IARC Scientific Publications No. 97

International Agency for Research on Cancer

Lyon, 1989

Published by the International Agency for Research on Cancer,
150 cours Albert Thomas, 69372 Lyon Cedex 08, France

© International Agency for Research on Cancer, 1990

Distributed by Oxford University Press, Walton Street, Oxford OX2 6DP, UK

Distributed in the USA by Oxford University Press, New York

All rights reserved. No part of this publication may be reproduced, stored in a retrieval system, or transmitted, in any form or by any means, electronic, mechanical, photocopying, recording, or otherwise, without the prior permission of the copyright holder.

ISBN 92 832 1197 9

ISSN 0300-5085

Printed in the United Kingdom

Contents

Foreword vii

Epidemiological aspects of the relationship between exposure to silica dust and lung cancer
    L. Simonato and R. Saracci 1

Occuptional groups potentially exposed to silica dust: a comparative analysis of cancer mortality and incidence based on the Nordic occupational mortality and cancer incidence registers
    E. Lynge, K. Kurppa, L. Kristofersen, H. Malker and H. Sauli 7

A case-referent study on lung cancer mortality among ceramic workers
    S. Lagorio, F. Forastiere, P. Michelozzi, F. Cavariani, C.A. Perucci and O. Axelson 21

Silica and cancer associations from a multicancer occupational exposure case-referent study
    J. Siemiatycki, M. Gérin, R. Dewar, R. Lakhani, D. Begin and L. Richardson 29

Cancer mortality of granite workers 1940–1985
    R.S. Koskela, M. Klockars, E. Järvinen, A. Rossi and P.J. Kolari .. 43

A mortality study of a cohort of slate quarry workers in the German Democratic Republic
    W.H. Mehnert, W. Staneczek, M. Möhner, G. Konetzke, W. Müller, W. Ahlendorf, B. Beck, R. Winkelmann and L. Simonato 55

Occupational dust exposure and cancer mortality – results of a prospective cohort study
    M. Neuberger and M. Kundi 65

Lung cancer mortality among pottery workers in the United States
    T.L. Thomas 75

A mortality follow-up study of pottery workers: preliminary findings on lung cancer
    P.D. Winter, M.J. Gardner, A.C. Fletcher and R.D. Jones 83

Lung cancer risk among pneumoconiosis patients in Japan, with special reference to silicotics
  K. Chiyotani, K. Saito, T. Okubo and K. Takahashi 95

Mortality from specific causes among silicotic subjects: a historical prospective study
  F. Merlo, M. Doria, L. Fontana, M. Ceppi, E. Chesi and L. Santi 105

Lung cancer incidence among Swedish ceramic workers with silicosis
  G. Tornling, C. Hogstedt and P. Westerholm 113

Index 121

Foreword

Silica dust has long been recognized as a major occupational hazard, causing disability and deaths worldwide among workers in several industries such as mining and quarrying, refractory materials industries, potteries and foundries. The profound and often extensive lesions caused by exposure to crystalline silica in the lung led naturally to the question of whether these lesions also favour the occurrence of lung cancer. A number of early studies, mostly based on autopsy material, gave a negative answer to this question and the view that exposure to silica does not increase the risk of lung cancer became widely accepted.

It is only in the last decade that this issue has been taken up again, using more powerful epidemiological methodology than in the previous investigations. As a result, associations are now documented between work involving exposure to silica dust and cancer of the lung. To contribute to the clarification of these findings, the IARC convened in June 1986 a working group to evaluate the carcinogenic risk of silica dust within its programme of Monographs on the Evaluation of Carcinogenic Risks to Humans (IARC Monographs Vol. 42, *Silica and Some Silicates*, 1987). At the same time, the Agency activated a study group with the task of producing new epidemiological information on some of the less investigated aspects of the lung cancer–silica dust relationship. The results of these studies are now assembled in the present volume, adding to the evidence of a link between silica dust and lung cancer and at the same time bringing into better focus a number of still unresolved questions, notably on the relative role of silica and of silicates in the occurrence of lung cancer. As silica dust exposure continues to occur extensively in both developed and developing countries, it is to be hoped that these results, in addition to having intrinsic scientific interest, will encourage further research that will provide definitive answers to the outstanding questions relevant to the occupational health significance of silica.

L. Tomatis, M.D.
Director, IARC

Epidemiological aspects of the relationship between exposure to silica dust and lung cancer

L. Simonato[1] and R. Saracci[2]

[1]Centre of Environmental Carcinogenesis,
University of Padua, Padua, Italy
[2]Unit of Analytical Epidemiology, International Agency for
Research on Cancer, Lyon, France

Introduction

The question of whether exposure to silica, besides its well known health effects on the respiratory system, also represents a carcinogenic hazard for humans has been somewhat quiescent for many decades. Most occupational health experts have been reassured by the lack of association between silicosis and lung cancer in post-mortem series. Yet even in the first volume of the *IARC Monographs* series, silica was quoted as one of the possible causes of the lung cancer excess found in iron ore miners (IARC, 1972).

Since David Goldsmith drew the attention of the scientific community to this issue (Goldsmith, 1982), a relatively large number of studies have been conducted aimed at investigating the possible carcinogenic role of silica. The International Agency for Research on Cancer (IARC) evaluated the experimental and human evidence in 1987 and concluded that there was sufficient, for the former, and limited, for the latter, evidence of carcinogenicity of crystalline silica. This evaluation was mainly based on the most recent studies and is further stimulating the debate on hazards associated with silica. It is likely to lead to the setting up of more studies in order to provide a definite assessment of the carcinogenicity of this substance.

The IARC in 1983 assembled a group of research teams with the goal of coordinating studies to improve understanding of the potential carcinogenic risk from exposure to silica, with particular attention being paid to exposure to silica in the working environment in the absence of other agents known to be carcinogenic for the lung. The present volume brings together the results of the studies carried out within this project, and thus provides a concise picture of the current state of our knowledge. This introductory chapter should be regarded neither as an exhaustive overview nor as a re-evaluation of the evidence of the carcinogenicity of silica. Our only aim has been to focus on the main epidemiological aspects of the issue and to draw attention to the key questions which, in our opinion, have not yet been satisfactorily answered.

Studies and interpretations

Three main categories of workers have been studied to provide epidemiological evidence on the carcinogenicity of silica dust: (*a*) workers compensated for silicosis; (b) workers exposed to silica in industries whose working environment is known to be contaminated with carcinogenic agents; and (c) workers exposed to silica in industries with little or no contamination by known lung carcinogens.

Studies on workers compensated for silicosis

Leaving aside for the moment the interpretation of the findings in causal terms, it is interesting to notice how the interpretation of a long established and commonly accepted human observation has changed rather dramatically in the last few years from negative to positive. In Volume 1 of the *IARC Monographs* previously quoted, the following statement may be found '... there is much evidence that silicosis *per se* does not predispose to and is not associated with development of primary cancer of the lung...' (IARC, 1972). However, cohort studies of workers compensated for silicosis presented in this book or published elsewhere have consistently reported an excess of lung cancer of two-fold or greater (Westerholm, 1980; Finkelstein *et al.*, 1982; Kurppa *et al.*, 1986; Zambon *et al.*, 1987; Chiyotani, Merlo, Tornling, this volume). This evidence has been supported by case–control studies (Mastrangelo *et al.*, 1988; Forastiere *et al.*, 1989; Lagorio, this volume), with the exception of the study by Hessel and Sluis-Cremer (1986).

While there does not seem to be much doubt about this association, important limitations still prevent a full understanding of these findings, but some interesting hypotheses have been advanced, on which epidemiological investigations are warranted. One of the remarks often made is that silicotics are a highly selected population and that, consequently, it is very difficult to choose a suitable referent population. It should be noted that this bias can be introduced only if subjects exposed to silica and suffering from lung cancer have more chance of being compensated for silicosis than those not affected by lung cancer. This could well be the case in some countries, but has not so far been demonstrated in any of the studies. More important could be the bias introduced when the source of the study population is a hospital rather than, for example, a national insurance scheme. In this case it can be suspected that silicotics with lung cancer are more likely to be admitted to hospital than silicotics not affected by lung cancer.

Although it was not the main goal of the teams reporting here to contribute further evidence on the association between lung silicosis and lung cancer, the interpretation of this relationship remains crucial and will be reconsidered in the final part of this chapter.

Studies on workers exposed to silica and to known occupational carcinogens

Most frequently, and for the largest proportion of subjects, exposure to silica occurs with concomitant exposure to other carcinogens. Most, if not all, mining activities involve at least some exposure to silica in addition to the well known exposure to radon, while foundry workers exposed to various polycyclic aromatic hydrocarbons may also suffer from exposure to silica.

Given the difficulties in assessing not only the level of exposure to silica but also exposure to other established carcinogenic agents, the epidemiological studies carried out on these working populations provide a limited contribution to the understanding of the possible carcinogenic risk from exposure to silica. When silica exposure has been specifically investigated (Fletcher, 1985), the results have not indicated a clear role of silica in increasing lung cancer risk.

Studies on workers exposed to silica with little or no contamination by known lung carcinogens

To study occupational populations exposed to silica with a low risk of confounding effects from known carcinogens was the main target of this project. The chapters in this volume by Koskela et al., Lagorio et al., Mehnert et al., Thomas and Winter et al. address this issue, while in a sixth chapter reporting the case–control study by Siemiatycki et al., a number of potential occupational confounders have been controlled for.

Table 1 summarizes the main findings of the studies specific to a given industry in which exposure to silica occurs. All these studies report a modest excess of lung cancer which, although it does not exceed a two-fold increase, is statistically significant in four out of the five studies.

Table 1. Studies on workers exposed to silica in industries with little or no contamination with other lung carcinogens: main findings reported in this volume[a]

Study population	Overall rate ratio	Time since first exposure	Duration of exposure	Estimated dose
Granite workers (Koskela et al.)	1.56*	+	n.a.	n.a.
Slate quarry workers (Mehnert, 1989)	1.1 1.8 silicotics 0.9 non-silicotics	+ + +	+	–
Ceramic workers (Lagorio et al.)	2.0* 3.9* silicotics 1.4 non-silicotics	n.a.	±	n.a.
Pottery workers (Thomas)	1.2* 1.8* sanitary ware	–	–	+
Pottery workers (Winter et al.)	1.3*	n.a.	n.a.	±

*, statistically significant at 0.05 level; +, positive association; –, negative association; ±, doubtful association; n.a., not available

It is interesting to notice that both of the two studies investigating silicotics and non-silicotics separately have found that the excess was due mainly to the lung cancer pattern among silicotics.

The relationships with latency, duration of exposure and qualitative or quantitative estimates of exposure are not consistent across studies.

The case–control study by Siemiatycki et al. (1989), which is not reported in the table, also finds a moderate increase in the relative risk, which is supported by the results of the analysis using dose estimates. Only lung cancer shows an increased risk in this multi-site multi-exposure case–control study in which a number of occupational and non-occupational potential confounders are adjusted for.

Overall, these studies are suggestive of a moderate excess lung cancer risk among workers exposed to silica which, in the studies that collected data on smoking, could not be ascribed to this personal habit. Exposure to radiation did not occur in the occupational situations investigated. Some exposure to carcinogens such as asbestos or polycyclic aromatic hydrocarbons cannot be fully excluded, but their confounding effect, if any, is unlikely to be strong.

Conclusions

Our present knowledge of the potential carcinogenic effect of silica dust is characterized by three main findings: (i) silica is carcinogenic in experimental systems (IARC, 1987); (ii) lung cancer risk is increased among workers exposed to silica and not exclusively among those exposed to known carcinogens; (iii) when investigated separately, the lung cancer risk is concentrated among the subpopulation of exposed workers who develop silicosis.

Researchers in cancer etiology here face an unusual situation in which a 'probable' carcinogen (IARC, 1987) appears to act mainly, if not exclusively, among exposed subjects affected by another respiratory disease caused by the same agent. There are serious difficulties for epidemiological research in trying to disentangle such a complex situation which, for the sake of simplicity, could be explained by either of two hypotheses: (*a*) exposure to silica is the direct cause of both silicosis and lung cancer and the concentration of the excess risk among silicotics is the obvious consequence of higher exposure. This explanation, although having the advantage of being simple, would imply a full parallelism of the two pathological processes, which would be a rather unusual situation; (*b*) exposure to silica is the indirect cause of the increased lung cancer risk through the development of silicosis, a pathological condition which could enhance the effects of exposure to other carcinogens, including or excluding silica (whether silica is itself active as a carcinogen at this stage is irrelevant for the definition of the model). According to this hypothesis the pathological condition of silicosis would have a promoting effect on the carcinogenic process.

Neither hypothesis has been fully explored and the contributions in this volume represent a first effort in this direction which, we hope, has contributed to the debate on this issue, which goes back a long time in the history of occupational medicine.

References

Finkelstein, M., Kusiak, R. & Suranyi, G. (1982) Mortality among miners receiving workmen's compensation for silicosis in Ontario: 1940-1975. *J. Occup. Med.*, 24, 663-667

Fletcher, A.C. (1986) The mortality of foundry workers in UK. In: Goldsmith, D.F., Winn, D.M. & Shy, C.M., eds, *Silica, Silicosis and Cancer: Controversy in Occupational Medicine* (Cancer Research Monographs, Vol. 2), New York, NY, Praeger, pp. 385-401

Forastiere, F., Lagorio, S., Michelozzi, P., Perucci, C.A. & Axelson, O. (1989) Mortality pattern of silicotics in Latium Region - Italy. *Br. J. Ind. Med.* (in press)

Goldsmith, D.F., Guidotti, T.L. & Johnston, D.R. (1982) Does occupational exposure to silica cause lung cancer? *Am. J. Ind. Med.*, 3, 423-440

Hessel, P.A. & Sluis-Cremer, G.K. (1986) Case-control study of lung cancer and silicosis. In: Goldsmith, D.F., Winn, D.M. & Shy, C.M., eds, *Silica, Silicosis and Cancer: Controversy in Occupational Medicine* (Cancer Research Monographs, Vol. 2), New York, NY, Praeger, pp. 351-355

IARC (1972) *IARC Monographs on the Evaluation of the Carcinogenic Risk of Chemicals to Man*, Vol. 1, Lyon, International Agency for Research on Cancer, pp. 29-39

IARC (1987) *IARC Monographs on the Evaluation of the Carcinogenic Risk of Chemicals to Humans*, Vol. 42, *Silica and Some Silicates*, Lyon, International Agency for Research on Cancer

Kurppa, K., Gudbergsson, H., Hannunkari, I., Koskinen, H., Hernberg, S., Koskela, R.-S. & Ahlman, K. (1986) Lung cancer among silicotics in Finland. In: Goldsmith, D.F., Winn, D.M. & Shy, C.M., eds, *Silica, Silicosis and Cancer: Controversy in Occupational Medicine* (Cancer Research Monographs, Vol. 2), New York, NY, Praeger, pp. 311-319

Mastrangelo, G., Zambon, P., Simonato, L. & Rizzi, P. (1988) A case-referent study investigating the relationship between exposure to silica dust and lung cancer. *Int. Arch. Occup. Environ. Health*, *60*, 299-302

Westerholm, P. (1980) Silicosis. Observation on a case register. *Scand. J. Work Environ. Health*, *6*, (Suppl. 2), 1-86

Zambon, P., Simonato, L., Mastrangelo, G., Winkelmann, R., Saia, B. & Crepet, M. (1987) Mortality of workers compensated for silicosis during the period 1959-1963 in the Veneto Region of Italy. *Scand. J. Work Environ. Health*, *13*, 118-123

Occupational groups potentially exposed to silica dust: a comparative analysis of cancer mortality and incidence based on the Nordic occupational mortality and cancer incidence registers

E. Lynge[1], K. Kurppa[2], L. Kristofersen[3], H. Malker[4] and H. Sauli[5]

[1]*Danish Cancer Registry, Copenhagen, Denmark*
[2]*Institute of Occupational Health, Helsinki, Finland*
[3]*Central Bureau of Statistics, Oslo, Norway*
[4]*National Board of Labour Protection, Solna, Sweden*
[5]*Central Bureau of Statistics, Helsinki, Finland*

Summary. We have analysed mortality and cancer incidence data available in census-based record-linkage studies from the Nordic countries for males in occupational groups with potential exposure to silica dust. The study showed an excess lung cancer risk for foundry workers in all the Nordic countries, and also for miners in Sweden. These results are consistent with the findings of previous in-depth epidemiological studies. The lung cancer risk did not differ significantly from that of the respective national populations for males working in glass, porcelain, ceramics and tile manufacture, in excavation, and in stone quarries, sand and gravel pits. Stone cutters, who are probably not exposed to known lung carcinogens at the workplace but in some places to high concentrations of silica dust, showed a significant excess lung cancer risk in both Finland and Denmark.

Introduction

Few in-depth studies on lung cancer in relation to exposure to silica dust have been carried out. We have therefore considered it useful to explore the general picture of lung cancer occurrence in occupational groups with potential exposure to silica dust by using data from the Nordic occupational mortality and cancer incidence registers. Silica dust exposure has been an issue of concern within occupational health in the Nordic countries during the last 50 years (Ahlmark, 1967).

The Nordic registers are all based on census data. Information on occupation therefore refers to one point in time only, and the classifications are often too broad to enable workers with specific exposures to be identified. For example, based on the occupation codes, it is possible to identify 'foundry workers', but not to distinguish between moulders, core-makers, casters, etc. Similarly, based on the industry codes, it is not possible to identify small industries where workers are exposed to high levels of silica dust, such as quartz mills and scouring powder factories. With these limitations in mind, we decided to consider data for foundry workers and miners, as both of these groups have been the subject of previous in-depth studies. They thus offered a basis for validation of the register data. From foundry workers and miners, we moved on to study other occupational groups potentially exposed to silica dust.

Materials and methods

In the Nordic census-based occupational mortality or cancer incidence registers, the cross-sectionally registered census populations are followed up for mortality or cancer incidence by individual record linkage. Record linkage for the entire national population is made possible by the existence of personal identification numbers. In all tabulations, occupation and industry refer to the status at the time of the census, and the analysis was restricted to males who were then 20-64 years old, as information on previous occupation is not available for pensioners in the censuses. In all four countries, industry has been coded according to modified versions of the International Standard Industrial Classification (ISIC). In Norway (Statistisk Sentralbyrå, 1975), Sweden (Statistiska Centralbyrån, 1964) and Finland (Tilastokeskus, 1974) occupation has been coded according to modified versions of the International Standard Classification of Occupation (ISCO), whereas a special code was used in Denmark (Danmarks Statistik, 1974). In the Danish code, a distinction is made between self-employed, skilled and unskilled workers, whereas this is not the case in ISCO.

Based on the codes for occupation and industry, an attempt was made to identify groups of persons potentially exposed to silica dust. The following job categories were selected for analysis: males in foundry work, mining, glass, porcelain, ceramics and tile manufacture, excavation, work in quarries, sand and gravel pits, and stone cutting. The relevant industry and occupation codes are given in the various tables. The four registers used are described below.

Norway

The 1970 census population was followed up for mortality over the ten-year period 1 November 1970-31 October 1980. All deaths during the follow-up period were coded according to the Eighth Revision of the *International Classification of Diseases*; lung cancer was coded 162. Data on possible cause of death and person-years at risk, taking death and emigration into account, were recorded for each of the 2 123 748 persons aged 20-64 at the time of the census. The expected number of lung cancer deaths was calculated by multiplying the person-years at risk, for each of the five-year birth cohorts of silica-dust-exposed males, by the mortality rate for the equivalent five-year birth cohorts of all males who were economically active at the time of the census. As an index of the relative mortality of silica-dust-exposed males aged 20-64 at the time of the census, the observed number of lung cancer deaths was divided by the expected number. This

Sweden

The 1960 census population was followed up for cancer registrations during the 19-year period 1 January 1961–31 December 1979. All cancer cases were coded according to the Seventh Revision of the *International Classification of Diseases*; lung cancer was coded 162.0–1. Person-years at risk were not calculated in Sweden. Instead, 'incidence rates' were calculated by dividing the number of cancer cases in five-year birth cohorts in the follow-up period by the number of persons in the equivalent five-year age groups at the time of the census. The expected number of lung cancer cases was calculated by multiplying the number of silica-dust-exposed males in a given region of Sweden in each five-year age group at the time of the census by the 'incidence rate' for the equivalent five-year birth cohort of all males in the region. National figures were obtained by aggregating both observed and expected numbers across the 27 Swedish regions. As an index of the relative 'incidence' among the silica-dust-exposed males aged 20–64 at the time of the census, the observed number of lung cancer cases was divided by the expected number (Malker & Weiner, 1984).

Finland

As in Norway, the 1970 census population was followed up for mortality for the ten-year period 1 January 1971–31 December 1980. All deaths during the follow-up period were coded according to the Eighth Revision of the *International Classification of Diseases*; lung cancer was coded 162. The follow-up procedure was similar to that used in Norway, and possible cause of death and person-years at risk were recorded for each person aged 20–64 at the time of the census. Expected numbers of lung cancer deaths and SMRs for silica-dust exposed males aged 20–64 at the time of the census, were calculated in the same way as described for Norway (Sauli, 1979, and personal communication).

Denmark

The 1970 census population was followed up for cancer incidence for the ten-year period 9 November 1971–8 November 1980. Cancer cases were coded according to the Seventh Revision of the *International Classification of Diseases*; lung cancer was coded 162.0–1. The follow-up procedure was similar to that used in Norway. For each person aged 20–64 at the time of the census, possible lung cancer diagnosis and person-years at risk were recorded taking death, emigration and lung cancer morbidity into account. The expected number of lung cancer cases was calculated by multiplying the person-years at risk for each of the five-year birth cohorts of silica-dust-exposed males by the incidence rate for the equivalent five-year birth cohorts of all males who were economically active at the time of the census. As an index of the relative incidence among silica-dust-exposed males aged 20–64 at the time of the census, the observed number of lung cancer cases was divided by the expected number. This index is equivalent to a standardized incidence ratio (Lynge, 1982).

Indices of relative mortality

As explained above, three different types of indices of relative mortality or incidence have been used in the Nordic census-based studies. However, these indices can all be considered as estimates of the relative risk of lung cancer of silica-dust-exposed males. These indices are therefore referred to in this study as relative risks (RR). Confidence intervals (95% CI) were calculated on the assumption that the observed numbers of deaths/cancer cases followed a Poisson distribution.

Discussion

Foundry work

Table 1 shows the RR for lung cancer among foundry workers. A statistically significant excess risk was observed in Norway (RR=1.73), Sweden (RR=1.38) and Finland (RR=1.56), and a statistically non-significant excess risk was seen among skilled moulders in Denmark (RR=1.60).

These results are in line with those of previous studies of foundry workers in Finland (RR=1.51) (Koskela et al., 1976), steel foundry workers in Canada (RR=2.50) (Gibson et al., 1977), gray iron foundry workers in the USA (RR=1.26) (Decouflé & Wood, 1979), members of the Molders and Allied Workers Union in the USA (RR=1.44) (Egan-Baum et al., 1981), foundry workers in the Pennsylvania steel industry (RR=7.1) (Blot et al., 1983), and English steel foundry workers, (RR=1.42) (Fletcher & Ades, 1984). The fact that only the skilled moulders showed an excess risk in Denmark might reflect a high turnover among unskilled foundry workers in this country. A previous Danish mortality follow-up study of participants in foundry health surveys showed a RR for lung cancer of 1.15, with a significant excess only for those who reported more than 25 years of foundry work (Sherson & Iversen, 1986). Foundries in the Nordic countries are mainly iron foundries where the only known carcinogens to which workers may be exposed are polycyclic aromatic hydrocarbons (PAH). A weak association has previously been shown to exist between the level of PAH exposure in iron foundries and the risk of lung cancer (Tola et al., 1979).

Mining

Table 2 shows that Norwegian miners had a lung cancer risk close to the average, whereas an excess risk was seen in Sweden both for iron ore miners (RR=3.19) and other ore miners (RR=3.71). There is a non-significant excess risk for the small group of Finnish iron ore miners (RR=1.78) and a significant excess risk for non-ferrous ore miners (RR=5.02). Denmark has no mining industry.

Mining is associated with a complex exposure pattern that can vary with the minerals concerned, equipment, location, etc. The lung cancer risk in Norwegian miners, found in this study to be close to the average for the national population, is in line with the results of a previous study, which showed no excess lung cancer risk among employees of the iron mines in northern Norway (Saugstad, 1983).

However, a significant excess risk of lung cancer was found among Swedish miners in both iron ore and other ore mining. This is in line with previous observations. An excess lung cancer risk has been observed in studies of iron ore

Table 1. Relative risk of lung cancer among foundry workers[a]

Country, industry and industry code[b]	Occupation and code[b]	No. of persons	O	E	RR	95% CI
Norway						
All	(737) Metal foundry workers	3 911	25	14.46	1.73	1.12–2.25
Sweden						
All	(736) Foundry workers	10 893	126	91.43	1.38	1.16–1.64
Finland						
(371XX) Iron and steel basic industry	(63X) Smelting, metallurgical foundry	3 466	32	20.51	1.56	1.31–1.86
Denmark						
All	(365) Skilled moulders	805	13	8.10	1.60	0.85–2.74
All	(416) Unskilled foundry workers	1 731	10	17.00	0.59	0.28–1.08
(341) Iron foundry	(480) Unskilled workers NEC[c]	1 290	12	10.85	1.11	0.57–1.93

[a] O, observed; E, expected; RR, relative risk; 95% CI, confidence interval on the assumption of a Poisson distribution
[b] In parentheses
[c] Not elsewhere classified

Table 2. Relative risk of lung cancer among miners[a]

Country, industry and industry code[b]	Occupation and code[b]	No. of persons	O	E	RR	95% CI
Norway						
(121) Iron ore mining						
(122) Pyrite and copper ore mining						
(129) Metal mining not elsewhere classified	(5XX) Mining and quarrying work, etc.	950	5	3.68	1.36	1.44–3.17
	(5XX) Mining and quarrying work, etc.	1167	5	4.99	1.00	0.33–2.34
Sweden						
(101) Iron ore mining	(5XX) Mining and quarrying	5914	124	38.84	3.19	2.92–3.49
(102) Other ore mines	(5XX) Mining and quarrying	1407	31	8.35	3.71	3.10–4.44
Finland						
(2301X) Iron ore mining	(4XX) Mining and quarrying	342	2	1.12	1.78	0.22–6.45
(320X) Non-ferrous ore mining	(4XX) Mining and quarrying	1312	21	4.18	5.02	3.11–7.68

[a] O, observed; E, expected; RR, relative risk; 95% CI, 95% confidnce interval on the assumption of a Poisson distribution
[b] In parentheses

underground miners in Malmberget, northern Sweden (RR=3.42) (Radford & St Clair Renard, 1984), of iron ore miners in Grängesberg, southern central Sweden (RR=14.4) (Edling, 1982), of Swedish zinc and lead ore miners (RR=13) (Axelson et al., 1971), and of Swedish copper ore miners (RR=4.1) (Dahlgren, 1979). Control for smoking in the Malmberget study only reduced the expected number of lung cancer cases from 14.6 to 12.8 (Radford & St Clair Renard, 1984). Diesel drive equipment was not introduced before the 1960s. Iron oxide is unlikely to be the carcinogenic agent as no excess of lung cancer was found in non-miners chronically exposed to iron oxide dust (Axelson & Sjøberg, 1979). For all groups of Swedish miners, the excess lung cancer risk has been ascribed to the high levels of radon daughters in Swedish underground mines, and there is evidence of a positive dose-response pattern (Radford & St Clair Renard, 1984). The possible influence of silica dust exposure has been considered in only one of the Swedish studies of miners. A case-control analysis from Malmberget showed X-ray evidence of silicosis not to be more prevalent among lung cancer patients than among other underground miners (Radford & St Clair Renard, 1984). It is noteworthy that the Swedish follow-up study of mortality in silicotics showed a RR of lung cancer for silicotics from 'mining, quarrying, and tunnelling' of 4.29, 95% CI 2.9–6.1, whereas the RR for silicotics from other industries was 1.81, 95% CI 1.1–2.8 (Westerholm, 1980). Exposure to radon daughters in mining might be a possible explanation of this difference.

A significant excess risk of lung cancer was seen in Finnish non-ferrous ore miners, most of whom work underground in copper, nickel, zinc and chromium mines (Roman, 1979), and are thus exposed to radon daughters. The miners in our study did not work in the Finnish anthophyllite asbestos quarries, from which an increased lung cancer risk has been reported (Meurman et al., 1974).

Porcelain, ceramics, tiles

Table 3 shows that the lung cancer risk for Norwegian males in the glass, porcelain, ceramics and tile industry was at the borderline of statistical significance (RR=1.79). A slight, non-significant excess was seen in the equivalent group of Finnish males (RR=1.27), whereas the overall figures for Swedish and Danish males were close to the national averages. At a more detailed level, a statistically significant excess risk was seen for Danish skilled and unskilled glass-workers when the jobs were grouped together (RR=1.74).

A previous Swedish case-control study of a glass-producing community showed an excess lung cancer risk among glass-workers (RR=2.0) (Wingren & Axelson, 1985), and it was pointed out that they were exposed to known lung carcinogens, such as asbestos and arsenic. A cohort study of US pottery workers showed a significant risk (RR=1.81) for lung cancer in pottery workers with exposure to high levels of silica. The significant excess was, however, limited to those pottery workers who were also exposed to talc (Thomas & Stewart, 1987). An Italian case-control study gave RR=2.0 for lung cancer among workers in the ceramics industry after controlling for smoking; the excess was statistically significant only among silicotic workers (Forastiere et al., 1986).

Table 3. Relative risk of lung cancer among workers in the glass, porcelain, ceramics and tile industries[a]

Country, industry and industry code[b]	Occupation and code[b]	No. of persons	O	E	RR	95% CI
Norway						
All	(81X) Glass, ceramics and clay work	2 052	15	8.40	1.79	1.00–2.95
All	(811) Glass-furnace workers, etc.	580	2	2.31	0.87	0.11–3.13
All	(812) Modellers and formers, etc. (ceramics)	451	3	1.71	1.75	0.36–5.13
Sweden						
All	(81X) Glass, porcelain, ceramics and tile work	10 315	94	89.54	1.05	0.95–1.16
All	(811) Glass-makers	2 142	18	14.62	1.23	0.73–1.95
All	(812) Moulders (ceramic products)	1 675	11	14.48	0.76	0.38–1.36
Finland						
All	(71X) Glass, ceramics, clay	2 829	22	17.32	1.27	0.80–1.92
All	(710) Glass moulders	1 095	6	5.17	1.16	0.43–2.53
All	(711) Potters	737	4	5.41	0.74	0.20–1.89
Denmark						
–	All workers in the glass, porcelain, ceramics and tile industries	5 226	55	53.40	1.03	0.90–1.18
All	(356) Glass-makers	289	3	1.85	1.62	0.33–4.74
(331) Glass industry	(480) Unskilled workers NEC[c]	1 211	21	11.97	1.75	1.09–2.68
All	(357) Ceramist	388	3	3.09	0.97	0.20–2.84
(332) Porcelain industry	(480) Unskilled workers NEC[c]	710	8	7.70	1.04	0.45–2.05
(330) Tile industry	(480) Unskilled workers NEC[c]	2 628	20	28.79	0.69	0.42–1.07

[a] O, observed; E, expected; RR, relative risk; 95% CI, 95% confidence interval on the assumption of a Poisson distribution
[b] In parentheses

Excavation

Workers in excavation, stone quarries, sand and gravel pits and stone cutters are occupational groups of particular interest in the study of lung cancer in relation to silica dust as they have no occupational exposure to known lung carcinogens.

Table 4 shows that excavation work was associated with a statistically significant excess lung cancer risk in Finland (RR=1.97), a small non-significant excess risk in Norway (RR=1.39), and no excess risk in Sweden. As with mining, this job category is not represented in Denmark.

The mortality of workers employed in excavation work has not previously been studied. There are differences between the census classifications in the Nordic countries with regard to such work. The Finnish figures cover only males employed in excavation, whereas those for Norway and Sweden include some persons employed in other building activities. In the interpretation of the results, a high proportion of smokers among Finnish construction workers (Asp, 1984), and the possible high labour turnover in the industry should also be taken into account. Only an in-depth study of the industry will thus make it possible to compare working conditions and exposure patterns across the Nordic countries.

Stone quarries, sand and gravel pits and stone cutters

Table 5 shows that there was no systematic association between work in stone quarries, sand and gravel pits and lung cancer in the Nordic census populations. Table 6 shows that stone cutters in Finland (RR=1.75) and skilled (RR=2.10) and self-employed (RR=2.90) stone cutters in Denmark have an excess lung cancer risk where, however, only the latter is statistically significant. Stone cutters in Norway and Sweden had no excess lung cancer risk.

A previous follow-up study of Finnish granite workers showed an excess lung cancer risk (RR=2.21) when allowance was made for a 15-year latency period (Koskela *et al.*, 1987). A proportional mortality study of granite cutters in the USA also showed a significant excess risk (RR=1.19) (Steenland & Beaumont, 1986), and an equivalent study of Vermont granite workers gave RR=1.2 (Davis *et al.*, 1983). Males registered as stone cutters may have worked in the industry for varying periods so that the relatively large group of stone cutters in the Swedish 1960 census included persons employed only temporarily in the trade. The small group of stone cutters in the Danish 1970 census, however, included only skilled workers whose main task was the production of granite building material and tombstones. Stone cutting includes a number of different operations, such as cutting, sawing and polishing. Measurements made during the Swedish silicosis project in 1968–71 showed wide variations in the total dust concentration as between different operations and samples (Gerhardsson *et al.*, 1974). It is therefore not possible, based on the crude census data available here, to evaluate possible differences between exposure levels in the Nordic countries. The mortality data from Norway and Finland allow tabulations of deaths due to silicotuberculosis (coded 010 in the Eighth Revision of the *International Classification of Diseases*) and pneumoconiosis silicotica (coded 515) as primary cause of death. For the small group of Norwegian stone cutters, no deaths were observed from these causes, as compared with 0.02 expected. For the small group of Finnish stone

Table 4. Relative risk of lung cancer in excavation work[a]

Country, industry and industry code[b]	Occupation and code[b]	No. of persons	O	E	RR	95% CI
Norway						
(42X) Construction other than buiding construction	(5XX) Mining and quarrying work, etc.	2120	11	7.92	1.39	0.69–2.49
Sweden						
(400) House building						
(415) Highway and hydraulic construction	(5XX) Mining and quarrying	4092	42	41.71	1.01	0.87–1.18
Finland						
(522XX) Excavating and foundations						
(5231X) Construction of highways, etc.						
(529XX) Other construction	(4XX) Mining and quarrying	2354	29	14.72	1.97	1.32–2.83

[a] O, observed; E, expected; RR, relative risk; 95% CI, confidence interval on the assumption of a Poisson distribution
[b] In parentheses

Table 5. Relative risk of lung cancer in quarrying and work in sand and gravel pits[a]

Country, industry and industry code[b]	Occupation and code[b]	No. of persons	O	E	RR	95% CI
Norway						
(141) Stone quarrying	(5XX) Mining and quarrying work, etc.	1098	6	4.57	1.31	0.48–2.86
(142) Gravel and sand pits	(5XX) Mining and quarrying work, etc.	126	–	0.56	–	–
(15X) Mineral quarrying	(5XX) Mining and quarrying work, etc.	406	1	1.81	0.55	0.01–3.08
Sweden						
(104) Quarries	(5XX) Mining and quarrying	1800	12	16.70	0.72	0.37–1.26
(105) Sand and clay pits	(5XX) Mining and quarrying	266	3	1.94	1.54	0.32–4.52
Finland						
(290111) Limestone quarrying	(4XX) Mining and quarrying	397	5	3.13	1.60	0.52–3.73
(290119) Quartz stone quarrying	(4XX) Mining and quarrying	126	–	–	–	–
(29012) Clay, gravel, sand mining	(4XX) Mining and quarrying	110	1	0.92	1.09	0.03–6.05
Denmark						
(120) Stone quarrying Gravel and sand pits	(480) Unskilled workers NEC[c]	983	11	10.86	1.01	0.51–1.81

[a] O, observed; E, expected; RR, relative risk; 95% CI, 95% confidence interval on the assumption of a Poisson distribution
[b] In parentheses
[c] NEC, not elsewhere classified

Table 6. Relative risk of lung cancer in stone cutting[a]

Country, industry and industry code[b]	Occupation and code[b]	No. of persons	O	E	RR	95% CI
Norway						
All	(857) Stone cutters and carvers	781	3	3.60	0.83	0.17–2.44
Sweden						
All	(856) Stone cutters	3275	37	37.85	0.98	0.83–1.16
Finland						
All	(756) Stone cutters	820	15	8.57	1.75	0.98–2.89
Denmark						
All	(358) Stone cutters	242	6	2.86	2.10	0.77–4.57
(336) Stone cutting	(021–023) Self-employed	158	7	2.41	2.90	1.17–5.98
(336) Stone cutting	(480) Unskilled workers NEC[c]	132	0	1.28	–	–

[a] O, observed; E, expected; RR, relative risk; 95% CI, 95% confidence interval on the assumption of a Poisson distribution
[b] In parentheses
[c] NEC, not elsewhere classified

cutters, two deaths were observed whereas only 0.04 were expected, which may be an indication of high exposure to silica dust in this group.

Conclusions

Results of record-linkage studies based on census data should be interpreted with caution. The results of the present study were, however, in line with previous results for occupational groups for which in-depth epidemiological studies had been carried out, e.g., for foundry workers and for Swedish miners. This fact encourages us to suggest that the following groups in our study, for whom a previously unnoticed excess lung cancer risk has been shown to exist, deserve further epidemiological study: Finnish miners, Danish glass-workers, Finnish males in excavation work, and Finnish and Danish stone cutters. Studies of stone cutters should, in particular, be given high priority, as the group is probably not exposed to known lung carcinogens at the work place, but sometimes to high concentrations of silica dust.

Acknowledgements

We are indebted for comments on the study design and analysis to Arne Brusgård, Timo Hakulinen, Eystein Glattre, Birgitta Malker, Lorenzo Simonato, Antti Tossavainen, Ebba Wergeland and Peter Westerholm. The study was supported by the International Agency for Research on Cancer (02/1/4 DEB 83/26).

References

Ahlmark, A. (1967) *Silicosis in Sweden* [in Swedish], Stockholm, Arbetarskyddsstyrelsen (Studia Laboris et Salutis, Vol. 1)

Asp, S. (1984) Confounding by variable smoking habits in different occupational groups. *Scand. J. Work Environ. Health*, *10*, 325-326

Axelson, O., Josefsson, H., Rehn, M. & Sundell, L. (1971) Swedish pilot study of lung cancer in miners [in Swedish], *Läkartidn.*, *68*, 5687-5693

Axelson, O. & Sjöberg, A. (1979) Cancer incidence and exposure to iron oxide dust. *J. Occup. Med.*, *21*, 419-422

Blot, W.J., Brown, L.M., Pottern, L.M., Stone, B.J. & Fraumeni, J.F. (1983) Lung cancer among long-term steel workers. *Am. J. Epidemiol.*, *117*, 706-716

Borgan, J.K. & Kristofersen, L. (1985) *Occupation and mortality 1970-80. Documentation note* [in Norwegian], Oslo, Statistisk Sentralbyrå (Interne notater 85/10)

Dahlgren, E. (1979) Lung cancer, cardiovascular disease and smoking in a group of miners [in Swedish], *Läkartidn.*, *76*, 4811-4814

Danmarks Statistik (1974) *Population and housing census 9 November 1970. C.I. Occupation and trade. Statistical tables 1974.* VII [in Danish], Copenhagen

Davis, L.K., Wegman, D.H., Monson, R.R. & Froines, J. (1983) Mortality experience of Vermont granite workers. *Am. J. Ind. Med.*, *4*, 705-723

Decouflé, P. & Wood, D.J. (1979) Mortality patterns among workers in a gray iron foundry. *Am. J. Epidemiol.*, *109*, 667-675

Edling, C. (1982) Lung cancer and smoking in a group of iron ore miners. *Am. J. Ind. Med.*, *3*, 191-199

Egan-Baum, E., Miller, B.A. & Waxweiler, R.J. (1981) Lung cancer and other mortality patterns among foundrymen. *Scand. J. Work Environ. Health*, *7*, suppl. 4, 147-155

Fletcher, A.C. & Ades, A. (1984) Lung cancer mortality in a cohort of English foundry workers. *Scand. J. Work Environ. Health*, *10*, 7-16

Forastiere, F., Lagorio, S., Michelozzi, P., Cavariani, F., Arcá, M., Borgia, P., Perucci, C. & Axelson, O. (1986) Silica, silicosis and lung cancer among ceramic workers: a case-referent study. *Am. J. Ind. Med.*, *10*, 363-370

Gerhardsson, G., Engman, L., Andersson, A., Isaksson, G., Magnusson, E. & Sundquist, S. (1974) *Silicosis Project: Final Report. Part 2. Aim, scope and results* [in Swedish], Stockholm, Arbetarskyddsstyrelsen (Undersökningsrapport AMT 103/74-2)

Gibson, E.S., Martin, R.H. & Lockington, J.N. (1977) Lung cancer mortality in a steel foundry. *J. Occup. Med.*, *19*, 807-812

Koskela, R.S., Hernberg, S., Kärävä, R., Järvinen, E. & Nurminen, M. (1976) A mortality study of foundry workers. *Scand. J. Work Environ. Health*, 2, suppl. 1, 73-89

Koskela, R.S., Klockars, M., Järvinen, E., Kolari, P.J. & Rossi, A. (1987) Cancer mortality of granite workers. *Scand. J. Work Environ. Health*, *13*, 26-31

Lynge, E. (1982) The Danish occupational cancer study. In: *Prevention of Occupational Cancer – International Symposium*, Geneva, International Labour Office (Occupational Safety and Health Series No. 46), pp. 557-568

Malker, H. & Weiner, J. (1984) *The cancer environment register, an example of the use of register epidemiology in occupational health* [in Swedish], Solna, Arbetarskyddsstyrelsen (Arbete och Hälsa, Vol. 9)

Meurman, L.O., Kiviluoto, R. & Hakama, M. (1974) Mortality and morbidity among the working population of anthophyllite asbestos miners in Finland. *Br. J. Ind. Med.*, *31*, 105-112

Radford, E.P. & St Clair Renard, K.G. (1984) Lung cancer in Swedish iron miners exposed to low doses of radon daughters. *New Engl. J. Med.*, *310*, 1485-1494

Roman, G.H. (1979) *World Mines Register 1979-80*, San Francisco, CA, Miller Freeman Publications

Saugstad, L.F. (1983) Cancer and atmospheric pollution. *Nord. Counc. Arct. Med. Res. Rep.*, *35*, 53-61

Sauli, H. (1979) *Occupational mortality 1971-75* [in Finnish], Helsinki, Tilastokeskus (Studies No. 54)

Sherson, D. & Iversen, E. (1986) Mortality among foundry workers in Denmark due to cancer and respiratory and cardiovascular diseases. In: Goldsmith, D.F., Winn, D.M. & Shy, C.M., eds, *Silica, Silicosis and Cancer: Controversy in Occupational Medicine* (Cancer Research Monographs, Vol. 2), New York, NY, Praeger, pp. 403-414

Statistiska Centralbyrån (1964) *Population census 1 November 1960. IX. Industry, occupation, commuting, households and education for the whole country, by county, etc.* [in Swedish], Stockholm

Statistisk Sentralbyrå (1975) *Population and housing census 1970*. Vol. II. *Industry, occupation and hours of work, etc.* [in Norwegian] Oslo

Steenland, K. & Beaumont, J. (1986) A proportional mortality study of granite cutters. *Am. J. Ind. Med.*, *9*, 189-201

Thomas, L.T. & Stewart, P.A. (1987) Mortality from lung cancer and respiratory disease among pottery workers exposed to silica and talc. *Am. J. Epid.*, *125*, 35-43

Tilastokeskus (1974) *Population census 1970. Occupation and social status* [in Swedish], Helsinki (FOS VI.C:104, del IX)

Tola, S., Koskela, R.-S., Hernberg, S. & Järvinen, E. (1979) Lung cancer mortality among iron foundry workers. *J. Occup. Med.*, *21*, 753-760

Westerholm, P. (1980) Silicosis – observation on a case register. *Scand. J. Work Environ. Health*, *6*, suppl. 2, 1-86

Wingren, G. & Axelson, O. (1985) Mortality pattern in a glass producing area in SE Sweden. *Br. J. Ind. Med.*, *42*, 411-414

A case-referent study on lung cancer mortality among ceramic workers

S. Lagorio[1], F. Forastiere[1], P. Michelozzi[1], F. Cavariani[1], C.A. Perucci[1] and O. Axelson[2]

[1]*Epidemiological Unit, Latium Regional Health Authority, Rome, Italy*
[2]*Department of Occupational Medicine, University Hospital, Linköping, Sweden*

Summary. A case–referent study has been carried out to test the hypothesis that silica-exposed ceramic workers have an increased risk of lung cancer. Next-of-kin interviews were conducted for 72 lung cancer cases and 319 referents, all deceased, to collect work histories and smoking habits. The diagnosis of silicosis was ascertained by checking the individual files of cases of silicosis where compensation had been received. It was found that, after controlling for age, period of death and smoking, workers in the ceramic industry had a higher lung cancer risk than those in other occupations in which there was no exposure to silica (Mantel-Haenszel rate ratio=2.0; 95% confidence interval (CI) = 1.1–3.5). This increased risk was mainly due to a rate ratio of 3.9 (95% CI=1.8–8.3) for silicotic individuals, while for non-silicotic ceramic workers it was only 1.4 (95% CI=0.7–2.8). Exposure to other carcinogens in the workplace seems not to play any role in the development of lung cancer. Furthermore, the data do not suggest an increased risk for silicotic non-smokers. The results of the study tend to confirm previous evidence of an excess risk among silicotic subjects and points to a possible etiological role of the silicotic process itself in lung cancer.

Introduction

Experimental and epidemiological observations seem to indicate an etiological role of silica exposure in lung cancer induction (IARC, 1987). Follow-up studies of workers who received compensation for silicosis consistently show an excess of lung cancer mortality (Westerholm, 1980; Finkelstein *et al.*, 1982; Kurppa *et al.*, 1986; Zambon *et al.*, 1987). However, studies on silica-exposed subjects are controversial (Fletcher & Ades, 1984; Steenland & Beaumont, 1986; Koskela *et al.*, 1987), and it has been suggested that a confounding effect from concomitant exposures to other known carcinogens in the workplace could explain the excess risks that have been found (Heppleston, 1985).

Studies of workers with 'pure' exposure to silica dust, such as pottery workers employed in the manufacture of sanitary ware and crockery, might provide a better means of resolving the problem, but few such studies have been carried out on this occupational group. A cohort study of US pottery workers showed an increase in lung cancer mortality (Thomas & Stewart, 1987), especially among those exposed to non-fibrous talc in addition to high levels of silica. On the other hand, a census-based record-linkage study conducted in the Nordic countries did not show a clear excess risk for ceramic workers (Lynge et al., 1986).

We have therefore carried out a case–referent study to further test the hypothesis that ceramic workers might have an excess risk of lung cancer. The methods and some of the results have already been reported elsewhere (Forastiere et al., 1986); further information on the study is presented here.

Methods

The small town of Civitacastellana (15 606 inhabitants) in northern Latium, Italy, was chosen for a case–referent study in view of its long tradition of pottery manufacture and the relatively high proportion of the male population employed in the industry.

A total of 84 lung cancer (item 162, Eighth Revision of the *International Classification of Diseases*) cases, whose death occurred during the period 1968–84, was drawn from the local municipal registries of deaths together with 334 referents. The controls were initially chosen according to the following criteria: a case/control ratio of 1:4, age at death and year of death the same as those of the cases (± 5 years and ± 3 years respectively), and causes of death other than lung cancer and tumours of unspecified site.

To adequately represent the exposure frequency in the study base, deaths possibly due to diseases caused by the exposure of interest, namely silicosis and chronic bronchitis, were subsequently excluded from the referents and replaced by an equal number (35) of subjects with acceptable diagnoses. However, four referents certified as having died from cor pulmonale (item 426 in the Eighth Revision of the *International Classification of Diseases*) were retained in the control series.

The cases accepted for the study were reduced, after validation through hospital medical records, to 80; for 57 of them the diagnosis was confirmed clinically, mainly on the basis of radiological findings.

A next-of-kin interview was conducted to collect detailed work histories of the subjects as well as lifetime smoking habits. The presence of silicosis for which compensation had been received was assessed for both cases and referents by checking a list of the awards for pneumoconiosis released by the Italian Institute for Compensation for Occupational Diseases (INAIL). Exposure categories were established, 'ceramic workers' being considered as all those who had been employed as blue collar workers in a pottery factory for at least one year. They were subdivided into 'silicotics' (Sil+) and 'non-silicotics' (Sil−) based on the INAIL information. A separate category of 'quarrymen' was created, because of the potential exposure to silica dust, for those who had been working in a quarry in the area for at least one year but without ever having worked in the ceramics industry.

Exposure to silica dust is the main respiratory hazard in the ceramics industry. Nevertheless, other potential respiratory carcinogens present in the workplace must be taken into account, even though they have been used only in small amounts and the proportion of the workforce exposed to them has been quite low. Talc, possibly contaminated with asbestos fibres, has been used regularly in the slip-casting process to dust moulds; small amounts of chromate pigments are added to glazes and exposures to diesel exhaust fumes from the firing process may have occurred since the 1960s.

In order to evaluate a possible cluster of cases among subjects performing jobs entailing exposure to the above-mentioned potential carcinogens, all the work histories of the ceramic workers were reviewed to enable individuals to be classified according to a combination of job titles and departments. For each subject, the title of the job in the ceramic industry held for the longest period of time was used.

Finally, with regard to smoking habits, the subjects were divided into three categories according to the estimated daily number of cigarettes smoked: (1) non-smokers; (2) smokers of 1–20 cigarettes per day; and (3) smokers of more than 20 cigarettes per day; cigar and pipe smokers (16 referents) and cigarette smokers whose daily consumption was unknown (two cases and 17 referents) were included in the second category. All these classifications were made blindly with regard to the cause of death.

Mantel–Haenszel (MH) procedures (Mantel & Haenszel, 1959) were used to estimate overall rate ratios (MH–RR) equivalent to incidence density ratios (Miettinen, 1976; Axelson, 1983) and chi square values; standardized rate ratios and 95% confidence intervals (95% CI) were computed according to Miettinen (Miettinen, 1972 a,b, 1976). The overall homogeneity of the rate ratios using the logit approach was tested as suggested by Breslow and Day (Breslow & Day, 1980). The etiological fractions (EF) (population-attributable risks), namely the proportion of all cases due to the exposure to a given factor, were also calculated (Walter, 1978; Berrino, 1982).

Results

The results refer to 391 subjects, namely, 72 cases and 319 referents, because no next of kin could be traced for eight cases (10%) and 14 referents (4.4%), and for one additional referent the information about smoking was missing. Table 1 shows the distribution of cases and referents by exposure categories, together with rate ratios, 95% CI and etiological fractions. MH–RR were computed, taking into account, as potential confounders, age at death (< 65 years, > 64 years), period of death (1968–77, 1978–84) and smoking habits (non-smokers, 1–20 cigarettes per day, > 20 cigarettes per day).

For ceramic workers versus unexposed individuals, the MH–RR was 2.0 (95% CI=1.1–3.5); 'silicotics' (Sil+) showed a higher rate ratio (MH–RR=3.9; 95% CI=1.8–8.3) while for 'non-silicotic ceramic workers' (Sil–) a rate ratio of 1.4 (95% CI=0.7–2.8) was obtained. No excess was found for 'quarrymen' (MH–RR=1.0; 95% CI=0.4–2.4). The values of EF were 0.08 and 0.15 for 'non-silicotics' and 'silicotics', respectively.

Table 1. Lung cancer cases and referents by exposure category

Exposure	Cases	Referents	Crude rate ratio	SMR[a]	MH-RR[b] (point estimate)	95% CI[c]	EF[d]
Quarries	5	24	1.2	0.7	1.0	0.4–2.4	–
Ceramics							
Sil-[e]	18	79	1.3	1.3	1.4	0.7–2.8	0.08
Sil+[f]	15	25	3.4	3.7	3.9	1.8–8.3	0.15
Total	33	104	1.8	1.9	2.0	1.1–3.5	0.23

[a] Standardized mortality ratio, taking age, period of death and smoking into account (see text)
[b] Mantel-Haenszel rate ratio
[c] 95% confidence interval
[d] Etiological fraction
[e] Non-silicotic subjects
[f] Silicotic subjects

Rate ratios were also calculated taking into account both smoking and work in a ceramics factory (Sil+ and Sil−) as determinants of lung cancer; unexposed non-smokers were taken as the reference group (Table 2). Among non-smokers, only 'non-silicotics' had a slightly increased risk (MH–RR=1.9; 95% CI=0.4–10.2), while no cases were found among the 'silicotics'. For smokers, on the other hand, the relative risks were clearly increased for 'silicotics', as shown by the MH–RRs of 6.3 (95% CI=1.7–22.8) for moderate smokers (1–20 cigarettes per day) and 8.9 (95% CI=1.5–54.8) for heavy smokers (> 20 cigarettes per day).

The analysis of the job titles of ceramic workers was based on 32 cases and 90 referents since a complete description of the work history was missing for one case and 14 referents. No obvious cluster of cases was found in the jobs where the presence of other respiratory carcinogens was suspected, namely slip-casting (talc), firing (exhaust fumes) and glaze making/spraying (chromate pigments) (Table 3). However, the point estimates are clearly unstable because of the small numbers involved.

Discussion

The study showed that ceramic workers had a lung cancer risk twice that of the unexposed population. This increased risk applied mainly to silicotic subjects (MH–RR=3.9) while, for non-silicotic ceramic workers, the MH–RR was 1.4 and the confidence interval included unity. Furthermore, it should be noted that 23% of the cases in the population concerned could have been prevented, at least in theory, by eliminating occupational exposure to silica dust.

Table 2. Mantel-Haenszel rate ratios[a] and 95% CI[b] for silica exposure, silicosis and smoking habits[c]

Smoking (cigarettes per day)		Exposure		
		Unexposed	Ceramics	
			Sil-[d]	Sil+[e]
Non-smokers	Cases	4	3	0
	Referents	39	15	7
	MH-RR[f]	1.0	1.9	0
	95% CI	-	0.4-10.2	-
1-20	Cases	10	7	9
	Referents	94	46	13
	MH-RR[f]	1.0	1.4	6.3
	95% CI	0.3-3.4	0.4-5.3	1.7-22.8
20+	Cases	20	8	6
	Referents	58	18	5
	MH-RR[f]	3.7	3.6	8.9
	95% CI	1.3-10.8	0.9-14.8	1.5-54.8

[a] Taking age and period of death into account (see text)
[b] 95% confidence interval
[c] Unexposed non-smokers are taken as the reference group
[d] Non-silicotic subjects
[e] Silicotic subjects
[f] Mantel-Haenszel rate ratio

Table 3. Lung cancer cases and referents by exposure category based on combination of job title and department as obtained from work histories

Exposure	Cases	Referents	RR[a]	95% CI[b]
Unexposed	34	191	(1.0)	
Exposed to silica:				
Sanitary ware:				
Slip casting[c]	9	29	1.7	0.8-4.0
Firing[d]	10	26	2.2	1.0-4.8
Glaze making and spray glazing[e]	1	5	1.1	0.1-9.9
Unskilled labourers, repairers and cleaners	5	9	3.1	1.0-9.4
Other processes	2	4	2.8	0.5-15.0
Crockery manufacture	5	17	1.7	0.6-4.8

[a] Crude rate ratio
[b] 95% confidence interval
[c] Presumably exposed to talc
[d] Presumably exposed to diesel exhaust fumes
[e] Presumably exposed to chromate pigments

Information on individual exposure to dust is lacking; duration of employment was therefore used instead. There is no evidence of an increased lung cancer risk with prolonged exposure in either of the categories of ceramic workers (Table 4). However, in a comparison of the two 'ceramic' categories while also taking duration of employment into account, silicotic individuals still show a higher lung cancer risk ('ceramics Sil+' versus 'ceramics Sil-': MH-RR=3.1, 95% CI=1.2-7.5).

Table 4. Cancer risk of ceramic workers (cases and referents) by duration of employment (years)

Subjects		Duration of employment (years)		
		1-19	20-29	> 29
Sil-[a]	Cases	10	4	3
	Referents	34	18	20
RR[b]		1.0	0.8	0.5
95% CI[c]		-	0.2-2.8	0.1-2.0
Sil+[d]	Cases	3	4	8
	Referents	4	8	12
RR[b]		1.0	0.7	0.9
95% CI[c]		-	0.1-4.8	0.2-5.3

[a] Non-silicotic
[b] Crude rate ratio
[c] 95% confidence interval
[d] Silicotic

It has been suggested that smoking is an effect modifier (Goldsmith & Guidotti, 1986) of the association between silica exposure and lung cancer. In both a cohort study of silicotics (Zambon et al., 1987) and a case-referent study (L. Simonato, personal communication) conducted in northern Italy, an additive effect of smoking and exposure to silica in silicotics has been detected. The role of smoking as a co-factor is difficult to evaluate from our data because of the small numbers involved and the corresponding low power to detect an interaction. Furthermore, with regard to the effect of smoking *per se*, the true rate ratio seems to be underestimated in the two groups of unexposed moderate and heavy smokers, probably because of an overrepresentation of smokers among deceased controls (Gordis, 1982). The differences in the stratum-specific rate ratios with regard to smoking, however, are not statistically significant (test for homogeneity: $p > 0.05$) in either of the categories of ceramic workers so that a multiplicative model cannot be ruled out. Nevertheless, it is worth noting that no lung cancer excess was present among non-smoking silicotics; within the group of smokers, on the other hand, the rate ratios are suggestive of an underlying additive model.

The potential etiological role of other respiratory carcinogens within the ceramic industry cannot definitely be ruled out, but it seems less likely that, for example, talc is responsible for the increased risks found, as suggested by Thomas and Stewart (1987) in a cohort study of American pottery workers, since the rate ratio for slip casters is not particularly elevated. The relative risks found among those workers potentially exposed to chromate pigments and diesel exhaust fumes are also compatible with an overall rate ratio of 2.0 for ceramic workers.

Other problems of the internal validity of the study (including the possible confounding effect of other occupational exposures apart from the pottery environment) have already been discussed in a previous report (Forastiere et al., 1986).

Since silicosis and possibly also chronic bronchitis are associated with the exposure of interest, we excluded from the referents subjects who had died of these diseases in order to adequately represent the exposure frequency in the study base. In fact, if the aim is to achieve an unbiased representation of the population generating the cases, any disease entity can be included by using deceased referents as long as the force of mortality from that disease is similar among both the exposed and the unexposed subjects (i.e., the disease is not related to the factor under study) (Wang & Miettinen, 1982; Forastiere et al., 1987).

The inclusion of referents in the study whose deaths were caused by diseases mainly due to the exposure would have resulted in a false overestimation of the exposure in the study base and therefore a biased relative risk estimate. It may be argued that a bias might have been introduced in the opposite direction by the exclusions made if exposure to silica and silicosis decreased the chances of dying of other diseases, i.e., if silicosis and chronic bronchitis acted as competitive causes of death. This reasoning does not apply to an open study base and incidence density ratios as obtained in this study; even with a closed study base, the competition, if any, would theoretically apply proportionally to all other diseases and therefore also to lung cancer. Hence the relative risk estimates obtained after the exclusion of referents dying from competitive diseases are unbiased, irrespective of whether the study base is open or closed.

Finally, it is also possible that deaths from silicosis or chronic bronchitis might have replaced deaths from other causes differentially on the death certificates; i.e., more subjects who died from cardiovascular diseases might have been certified as dying from silicosis, whereas lung cancer cases might have been more correctly certified. This possibility cannot be entirely ruled out, but any such bias is likely to be minimal and would hardly be likely to lead to a wrong conclusion. We have shown, in fact, that even using a 'biased set' of referents consisting of the original group, without any exclusions, silicotic ceramic workers still have an increased lung cancer risk which is formally statistically significant (Forastiere et al., 1987).

In conclusion, the results from the present study are in line with previous findings of an excess lung cancer risk among workers exposed to silica. As suggested by follow-up studies, silicotic patients have a higher risk, and it could be argued that the silicotic process itself plays an etiological role in lung cancer development. The role played by smoking habits seems to be complex and should be further studied; our data tend to suggest, however, that there is no increased risk among non-smoking silicotics.

References

Axelson, O. (1983) Elucidation of some epidemiologic principles. *Scand. J. Work Environ. Health*, 9, 231-240

Berrino, F. (1982) Evaluation of the proportion of disease caused by occupational exposure. In: *Prevention of Occupational Cancer – International Symposium*, Geneva, International Labour Office (Occupational Safety and Health Series No. 46), pp. 291-295

Breslow, N.E. & Day, N.E. (1980) *Statistical Methods in Cancer Research*, Vol. 1, *The Analysis of Case-control Studies*. (IARC Scientific Publications No. 32), Lyon, International Agency for Research on Cancer, pp. 73-78

Finkelstein, M., Kusiak, R. & Suranyi, G. (1982) Mortality among miners receiving workmen's compensation for silicosis in Ontario: 1940-1975. *J. Occup. Med.*, 24, 663-667

Fletcher, A.C. & Ades, A. (1984) Lung cancer mortality in a cohort of English foundry workers. *Scand. J. Work Environ. Health*, 10, 7-16

Forastiere, F., Lagorio, S., Michelozzi, P., Cavariani, F., Arcá, M., Borgia, P., Perucci, C. & Axelson, O. (1986) Silica, silicosis and lung cancer among ceramic workers: a case-referent study. *Am. J. Ind. Med.*, 10, 363-370

Forastiere, F., Lagorio, S., Michelozzi, P., Cavariani, F., Arcá, M., Borgia, P., Perucci, C. & Axelson, O. (1987) Response to a letter. *Am. J. Ind. Med.*, 12, 221-222

Goldsmith, D.F. & Guidotti, T.L. (1986) Combined silica exposure and cigarette smoking: a likely synergistic effect. In: Goldsmith, D.F., Winn, D.M. & Shy, C.M., eds, *Silica, Silicosis and Cancer: Controversy in Occupational Medicine* (Cancer Research Monographs, Vol. 2), New York, NY, Praeger, pp. 451-459

Gordis, L. (1982) Should dead cases be matched to dead controls? *Am. J. Epidemiol.*, 115, 1-5

Heppleston, A.G. (1985) Silica, pneumoconiosis and carcinoma of the lung. *Am. J. Ind. Med.*, 7, 285-294

IARC (1987) *IARC Monographs on the Evaluation of the Carcinogenic Risk of Chemicals to Humans: Vol. 42, Silica and Some Silicates*, Lyon, International Agency for Research on Cancer

Koskela, R.-S., Klockars, M., Järvinen, E., Kolari, P.J. & Rossi, A. (1987) Cancer mortality of granite workers. *Scand. J. Work Environ. Health*, 13, 26-31

Kurppa, K., Gudbergsson, H., Hannunkari, I., Koskinen, H., Hernberg, S., Koskela, R.-S. & Ahlman, K. (1986) Lung cancer among silicotics in Finland. In: Goldsmith, D.F., Winn, D.M. & Shy, C.M., eds, *Silica, Silicosis and Cancer: Controversy in Occupational Medicine*, (Cancer Research Monographs Vol. 2), New York, NY, Praeger, pp. 311-320

Lynge, E., Kurppa, K., Kristofersen, L., Malker, H. & Sauli, H. (1986) Silica dust and lung cancer: results from the Nordic Occupational Mortality and Cancer Incidence Registry. *J. Natl Cancer Inst.*, 77, 883-889

Mantel, N. & Haenszel, W. (1959) Statistical aspects of the analysis of data from retrospective studies of disease. *J. Natl Cancer Inst.*, 22, 719-748

Miettinen, O.S. (1972a) Components of the crude risk ratio. *Am. J. Epidemiol.*, 96, 168-172

Miettinen, O.S. (1972b) Standardization of risk ratio. *Am. J. Epidemiol.*, 96, 383-388

Miettinen, O.S. (1976) Estimability and estimation in case-referent studies. *Am. J. epidemiol.*, 103, 226-235

Steenland, K. & Beaumont, J. (1986) A proportionate mortality study of granite cutters. *Am. J. Ind. Med.*, 9, 189-201

Thomas, T.L. & Stewart, P.A. (1987) Mortality from lung cancer and respiratory disease among pottery workers exposed to silica and talc. *Am. J. Epidemiol.*, 125, 35-43

Walter, S.D. (1978) Calculation of attributable risks from epidemiological data. *Int. J. Epidemiol.*, 7, 175-182

Wang, J.-D. & Miettinen, O.S. (1982) Occupational mortality studies. Principles of validity. *Scand. J. Work Environ. Health*, 8, 153-158

Westerholm, P. (1980) Silicosis: observations on a case register. *Scand. J. Work Environ. Health*, 6, suppl. 2, 1-86

Zambon, P., Simonato, L., Mastrangelo, G., Winkelmann, R., Saia, B. & Crepet, M. (1987) Mortality of workers compensated for silicosis during the period 1959-1963 in the Veneto region of Italy. *Scand. J. Work Environ. Health*, 13, 118-123

# Silica and cancer associations from a multicancer occupational exposure case-referent study

J. Siemiatycki[1,2], M. Gérin[3], R. Dewar[1,2], R. Lakhani[1], D. Begin[1] and L. Richardson[1]

[1]Centre de Recherche en Epidémiologie et Médecine Préventive,
Institut Armand-Frappier, Laval-des-Rapides, Québec, Canada
[2]Department of Epidemiology and Biostatistics,
McGill University, Montreal, Canada
[3]Département de Médecine du Travail et d'Hygiène du Milieu,
Université de Montréal, Montréal, Canada

Summary. A multicancer site, multifactor case-control study was undertaken to generate hypotheses about possible occupational carcinogens. Eligible cases, comprising all incident cases of cancer at 14 sites who were male, aged 35–70 and resident in Montreal, were subjected to probing interviews designed to obtain detailed lifetime job histories and information on potential confounders. Each job history was reviewed by a team of chemists who translated it into a history of occupational exposures. These occupational exposures were then analysed as potential risk factors in relation to the sites of cancer included. Over 3700 cases were interviewed and processed. For each site of cancer analysed as a case series, controls were selected from subjects with cancer at the other sites covered in the study. This report concerns the associations between silica exposure and each site of cancer. In initial screening analyses, there were significantly elevated odds ratios for silica and stomach cancer and for silica and non-adenocarcinoma lung cancer. More detailed analyses confirmed these associations, though there was no evidence of a dose-response relationship for stomach cancer. For lung cancer, there were significantly elevated risks in those with high-level long-term exposure. These results, which took into account several potential confounders including cigarette smoking and asbestos exposure, lend credence to the hypothesis that silica exposure increases the risk of lung cancer, and suggest the possibility of an effect on stomach cancer.

Introduction

There has been increasing controversy over the possible carcinogenicity of inhaled silica dust. While some studies have pointed to an excess risk of lung cancer, others have found no such risk (for reviews, see Goldsmith *et al.*, 1982; Heppleston, 1985; IARC, 1987). It is unclear whether these different findings are related to differences in quality or quantity of exposure to silica, or to methodological issues such as samples of inadequate size, insufficient follow-up, lack of control of important confounding factors such as smoking, and failure to use comparison groups or use of inappropriate ones. Some of these problems can be overcome in a case–referent study based on the general population. Furthermore, a case–referent study can address the problem of estimating risk due to silica exposure in the whole range of occupations and industries in which such exposure occurs, not just in the highly selected cohorts available for follow-up study. In the hope that such evidence could complement evidence from cohort studies, we have carried out a special analysis of detailed occupational histories collected in the context of a large-scale population-based case–referent study in Montreal with a view to determining cancer risks due to silica exposure. About 14 types of cancer among males are included in the study, each of which forms a case series for the purposes of analysis.

Methods

Multiexposure multisite monitoring study

The design briefly described here has been given in greater detail elsewhere (Siemiatycki *et al.*, 1981, 1987). A total of 14 sites of cancer were selected for study among males aged 35–70 resident in the area of Montreal. Case-ascertainment procedures were established in the 19 major hospitals, allowing us access to nearly all incident cases in the target population. Only if the diagnosis had been histologically confirmed was a case eligible for inclusion in the study.

The cases available for this analysis were incident from September 1979 to December 1985. However, because of limited resources, there were some interruptions in case ascertainment during this period. Each summer, for all sites, there was a 2–4-month ascertainment gap, depending on the interviewers' backlog from the previous year. Lung cancer was excluded in the second and third years, rectal cancer in the first and second year, and prostate cancer for part of the fourth year.

Active regular contact with hospital pathology departments provided rapid case notification. In the period covered by this report, 4576 subjects were found to be eligible for inclusion in the study. An interviewer visited the patient in hospital or at home, as required. Approximately half the cases were still in hospital when first contacted by the interviewer; the rest had been discharged or were diagnosed as outpatients. Completed interviews or questionnaires were obtained from 3726 subjects (81.4%). Reasons for non-response were: refusal – 8.1%; patient died, no next-of-kin found – 6.6%; patient discharged, no valid address available – 3.9%. Among the various types of cancer analysed as case series, response rates varied from 76.6 to 84.5%, with most between 80 and 82%. Face-to-face interview with the subject was the strategy of choice and 82% of completions were obtained in this way. However, telephone interviews and specially designed self-administered forms were used for hard-to-interview subjects and these provided 10 and 8%, respectively, of completions. Proxy information was sought from next-of-kin for subjects who had died; 19% of completions were obtained by this means.

The questionnaire was in two parts: (a) a structured section requesting information on important potential confounders; and (b) a semi-structured probing section designed to obtain a detailed description of each job the subject had had in his working lifetime. The interviewers were trained to probe for as much information as the patients could supply on the company's activities, the raw materials, final product and machines used, the man's responsibilities for machine maintenance, the type of room or building in which he worked, the activities of workmates around him, the presence of gases, fumes or dusts, and any other information which could furnish a clue as to the possible chemical or physical exposures incurred by the subject.

A team of chemists and hygienists working with us examined each completed questionnaire and translated each job into a list of potential exposures by means of a check-list form which explicitly lists some 300 of the most common occupational exposures, of which exposure to silica is one. For each product thought to be present in each job, the chemists noted the degree of confidence that the exposure had actually occurred (possible, probable, definite), the frequency of exposure during a normal working week (less than 5%, 5–30%, 30+%) and the level of concentration of the agent in the environment (low, medium, high). The degree of confidence depends on the clarity of the job descriptions obtained in interview and on the availability of documentary or other information on exposures in the type of workplace described. The frequency of exposure was usually readily available from the job description. The concentration or level of exposure was assessed with reference to a set of standards. For example, we defined low exposure to silica as that incurred by general workers on construction sites, medium exposure as that incurred by concrete finishers and high exposure as that incurred by moulders, core-makers and metal casters. Such benchmarks were defined for each substance on the check-list and each job was scored against these standards. The dates at which each job was started and ended were recorded, and thus the corresponding exposures in it. The jobs themselves were coded according to the *Canadian Classification and Dictionary of Occupations* (Ministry of Manpower and Immigration, 1974).

As a basis for this retrospective exposure assessment, the team of chemists relied on their own industrial experience and chemical knowledge, old and new technical and bibliographical material describing industrial processes, consultations with experts familiar with particular industries and, of course, previously coded files in the same job category. Neither interviewers nor chemists were aware of the patient's medical condition, thus eliminating one potential source of bias; a detailed description of the chemical coding is given in Gérin et al. (1985).

Restriction of study population

The criteria for patient inclusion have been listed above. One covariate which is a prime candidate as both a confounder and an effect modifier in the present context is ethnic group. The population of the Montreal area is made up of many ethnic groups. French Canadians, derived mainly from French settlers in the 17th and 18th centuries, account for two-thirds of the population, the other main groups being British, Italian, Jewish, Greek, German, Portuguese and other European. The melting-pot effect has been much less marked in Montreal than in other parts of North America, and the various groups have remained relatively homogeneous genetic and cultural entities. In addition, different ethnic groups in Montreal have tended to be concentrated in different occupational/industrial categories. In the past, French Canadians were under-represented in white collar jobs while certain ethnic groups (British, Jewish) were under-represented in blue collar ones. The opportunity for confounding is obvious, as is that for effect modification, since the different genetic profiles and cultural habits of the diverse ethnic groups may well interact with exposure in determining cancer risk. Any statistical approach aimed at testing and adjusting for confounding or effect modification by ethnic group on the cancer–exposure associations would require some unverifiable model assumptions and would not be totally efficient even if the model assumptions were correct. For instance, lumping together the many non-French ethnic groups (British, Italian, Jewish, middle European, etc.) may well leave residual confounding; keeping them separate would create very small numbers in some strata.

Because of the problems and uncertainties in this regard, we carried out the basic analyses twice, once among French Canadians only and once among all subjects, but controlling for ethnic group. The analyses among French Canadians have the advantage of eliminating one important source of confounding and, because they account for 62% of our study subjects, the loss of statistical precision is not too great. The analyses among all subjects have the advantage of greater numbers and thus greater statistical precision, and provide the opportunity to pick up risks in occupations in which ethnic groups other than French Canadians were concentrated.

Case groups and control groups

The monitoring study design calls for periodic analyses on cumulated subjects interviewed. Each series of subjects with a common tumour is compared with a referent series drawn from the other cancer sites included in the study. This strategy has been discussed elsewhere (Siemiatycki *et al.*, 1981).

For each cancer patient, the diagnosis, abstracted from medical records, was coded by topography and morphology according to the *International Classification of Diseases*. Cases could be grouped in many ways. For the most part we used the three-digit level of the *International Classification of Diseases* topography axis to define case series.

There were some exceptions, however, where the numbers of cases were so large as to make subdivision feasible. We analysed as case series each of the three topographic subcategories of colorectal cancer, namely colon (excluding sigmoid), sigmoid colon plus rectosigmoid junction, and rectum. We also analysed as case series each of four histological subcategories of lung cancer, namely oat-cell carcinoma, squamous-cell carcinoma, adenocarcinoma and other histological types (including unspecified morphology). The 78 cases diagnosed with primary tumours at two different sites have been included in both relevant case series.

For each case series, the control group consisted of all other sites in the study, with the following exceptions. Lung cancer was excluded from all control series because the very strong association between lung cancer and cigarette smoking is difficult to adjust for statistically; errors in measurement or in the functional form of the adjustment procedure could result in residual confounding. Also, for the three sites which were not ascertained in certain years of the study – lung, rectum and prostate – the controls consisted of other subjects who were ascertained in the same years as the corresponding cases so as to allow for possible variation in quality of interviewing or exposure assessment during the study period, despite our efforts to ensure consistency. Finally, we excluded from the control pool for a given site, any other site which is anatomically contiguous, both because there might be misclassification around the junction and because there is a greater likelihood of shared etiologies.

Table 1. Types of cancer analysed as case series, sites excluded from the 'cancer controls' series, and number of cases and 'cancer controls'

Cancer case series (ICD code)	Cancer sites excluded from control series[a]	All subjects		French Canadian subjects	
		No. of cases	No. of controls	No. of cases	No. of controls
Oesophagus (150)	Lung, stomach	107	2514	68	1436
Stomach (151)	Lung, oesophagus	250	2514	146	1436
Colon (153, except 153.3)	Lung, other colorectal	364	2080	196	1215
Rectosigmoid (153.3, 154.0)	Lung, other colorectal	233	2080	124	1215
Rectum (154, except 154.0)	Lung, other colorectal	190	1315	111	762
Liver (155)	Lung	50	2806	28	1613
Pancreas (157)	Lung	117	2741	66	1576
Lung: (162)					
Oat cell	Other lung	159	1523	112	875
Squamous cell	Other lung	359	1523	251	875
Adenocarcinoma	Other lung	162	1523	111	857
Other cell types[b]	Other lung	177	1523	116	875
Prostate (185)	Lung	452	1733	287	975
Bladder (188)	Lung, kidney	486	2196	278	1269
Kidney (189)	Lung, bladder	181	2196	96	1269
Melanoma of the skin (172)	Lung	121	2737	49	1594
Hodgkin's lymphoma (201)	Lung, other lymphoma	53	2599	127	1480
Non-Hodgkin lymphoma (200, 202)	Lung, Hodgkin's	206	2599	36	1480

[a] For each case series, all cancer patients interviewed served as referents with the exceptions listed in this column. Furthermore, for rectum, lung and prostate, only those subjects interviewed during the same ascertainment period as the three respective site series were used as referents.
[b] This is a heterogeneous grouping which includes large cell, spindle cell, adenosquamous and 'carcinoma, not otherwise specified'

Table 1 shows, for each of the sites analysed as case series, those that served as controls and the numbers of subjects included in the analyses.

Defining exposure groups

Exposure to silica was the factor of interest. For the purpose of computing odds ratios, exposure status could be dichotomized in a variety of ways. While any exposure/no exposure, as evaluated by the chemists, is the most straightforward, it runs the risk of diluting the 'exposed' with subjects who had had very low-level exposure or whose exposure status was questionable.

In the analyses outlined below, 'exposed' was sometimes defined as covering any exposure, of whatever degree, and sometimes according to a more stringent criterion based on degree and duration of exposure. For each substance to which a subject was thought to have been exposed, we had the following four types of information: (a) concentration of exposure; (b) frequency of exposure; (c) confidence that the exposure occurred; and (d) duration. Duration was expressed in years, while the other three factors were coded on three-point ordinal scales. To simplify the analyses, it was found useful to create a synthetic index of cumulative exposure based on concentration, frequency, confidence and duration. While each factor was simply coded 0, 1, 2 or 3, these ordinal values did not represent the relative weightings as the chemists actually used them. It was impossible to be precise, but we felt that the square of these terms more accurately reflected the way they were used. This can be seen most clearly in the case of the frequency scale, on which the absolute values corresponding to the three categories (< 5, 5–30, > 30%) are much steeper than 1:2:3. Although not as easily quantified as frequency, the difference between concentration levels of 3 and 1 would also reflect differences in actual values closer to nine-fold than to three-fold. Thus, before the exposure indices were created, the four point components were re-coded as 0, 1, 4 and 9.

For each year of the subject's career j, the average level of exposure to silica, X_j, was defined as:

$$X_j = \text{concentration}_j \times \text{frequency}_j \times \text{confidence}_j.$$

The cumulative exposure E to silica was defined as:

$$E = \Sigma_j X_j,$$

the sum over years of exposure.

Methods of analysis

The first stage in the analysis was an attempt to screen all cancer–silica associations so as to pick up those which warranted further attention. It is not immediately obvious whether it would be best to base this screening analysis on a dichotomization of any exposure versus no exposure or of substantial exposure versus no exposure; the choice depends on the numbers in each group and the shape of the dose–response curve, if there is an association. To maximize the chance of detecting true associations, we carried out the initial screening analysis first with exposure defined as any versus none and then as 'substantial' versus none. Substantial exposure to silica was defined as a value of E greater than the approximate median value of E among the exposed. Since, as explained above, the screening analyses were carried out both among all subjects and among French Canadians only, four sets of odds ratios were estimated for all cancer–silica associations.

For this purpose, we selected certain potential confounders to be included in every analysis, namely, ethnic group (except where the analysis was restricted to French Canadians), age, socioeconomic status as measured by self-reported income, cigarette smoking as measured by years of smoking times average daily amount, and blue/white collar status of the work history. This last variable was included because it is desirable that the case and control groups have similar blue/white collar job profiles, except for exposure to silica. The attempt to distinguish between 'clean' work histories and 'dirty' ones was based on a rough impressionistic evaluation by the team of chemists of the dirtiness of the job corresponding to each four-digit job category in the Canadian occupational classification system. Each job was scored from zero to six. Thus for each year of the man's career we derived a score associated with the job title held. The overall dirtiness score was an average of these annual scores over the man's working lifetime, and the variable was dichotomized at the median to create a blue/white collar variable.

The four sets of screening analyses were based on the procedure of Mantel and Haenszel (1959). The confounders were stratified as follows: ethnicity (French Canadian, other), age (35–54, 55–70), socioeconomic status as measured by income (below median, above median), dirtiness of job history (below median, above median), cigarette consumption (none, 0–600 cigarettes/day × years, 601+ cigarettes/day × years). A flexible programme was written for the purpose of computing large numbers of Mantel–Haenszel estimates in a single run (Dewar & Siemiatycki, 1985).

Associations with elevated odds ratios were selected for further in-depth investigation, with a view to taking into account not only the confounder variables included in the screening analyses, but also some specific occupational exposures to substances which were among those routinely coded. We carried out an intensive screening of all variables in our data set to detect those which were potential confounders in the sense that their inclusion in a model would change the odds ratio estimate by more than 10% (Siemiatycki et al., 1987). In addition to any variables flagged in this way, the asbestos variable was included in all in-depth analyses as a potential confounder. The analysis, which took all these confounders into account, was carried out by means of a logistic regression model using the GLIM programme (Baker & Nelder, 1978). Attributable risks were calculated according to the method of Bruzzi et al. (1986).

Results

Silica is one of the most common substances to which workers are occupationally exposed. Among all 3726 subjects, 25.0% were considered to have been exposed to silica at one time or another during their working lifetimes. Our team of chemists were certain of this exposure in 17.8% of the population. The high concentration of exposure was attributed to 1.9%, the high frequency of exposure to 10.4%, and 9.1% of the population had more than 20 years of exposure.

Numerically, the jobs most commonly represented among the 932 people exposed to silica were carpenters (10.2%), occupations in excavating, grading and paving (8.9%), mining and quarrying (6.1%), metal processing and related occupations (4.9%) and mechanics and repairmen (4.8%). The remaining 65% were distributed over a wide range of occupations and industries, about half of them in the construction industry.

As explained above, four estimates were made of each silica–cancer odds ratio, using a Mantel–Haenszel procedure. The results are shown in Table 2 for the 17 cancer types for which at least 40 subjects were interviewed, though for several cancer sites, the available numbers led to very wide confidence intervals. There are significantly or suggestively elevated odds ratios for stomach cancer and for three types of lung cancer, namely, oat-cell, squamous-cell and other, but not for adenocarcinoma of the lung. The lung cancer excess was apparent among French Canadians, the stomach cancer excess among all.

We then carried out in-depth analyses of two associations, namely, silica–stomach cancer and silica–lung cancer (except adenocarcinoma). These were carried out in the French Canadian subgroup as well as in the whole population. These analyses were based on the logistic regression model and included in the models the covariates included in the Mantel–Haenszel analyses plus the covariates flagged in the empirical search for confounders plus asbestos. For these analyses, the smoking confounder variable was taken as a six-category variable rather than the three-category one used in the Mantel–Haenszel analyses.

Dose–response relationships with silica exposure were studied by estimating the risk separately among each of the dimensions of exposure: duration, concentration, frequency and confidence. These dimensions were highly correlated amongst each other, and it proved impossible to adjust each for the effects of the other three.

Table 3 shows the results of these in-depth analyses among the French Canadian population. For stomach cancer there remains a hint of excess though the evidence is not convincing. Namely, there is no indication of increasing risk with increasing duration or concentration. The same analysis in the whole population showed somewhat higher odds ratios, though again there was no evidence of a dose–response relation. The results for lung cancer, by contrast, show a clear excess of risk, especially in the highest category of each exposure dimension.

We carried out some additional analyses in order to estimate the risk due to silica exposure in various occupations. For stomach cancer, the risk was particularly high for silica-exposed workers in excavation, grading and paving, who had an odds ratio of 2.1 as compared with subjects unexposed to silica. However, the odds ratio for everyone else exposed to silica was still somewhat elevated at 1.3. For lung cancer, the risk was also particularly high among silica-exposed excavation, grading and paving workers (odds ratio = 2.1). There was some lung cancer excess in many silica-exposed occupations.

Because of the importance of smoking in the etiology of lung cancer, it is of interest to examine the pattern of risk by silica and smoking. Table 4 shows the odds ratio for each smoking–silica combination defined by four smoking and three silica categories. Clearly, the effect due to smoking is much more important

Table 2. Four sets of odds ratios between exposure to silica dust and 17 types of cancer, based on Mantel–Haenszel analysis[a]

Site	All subjects Ever/never[b]			Substantial/never[c]			French Canadian subjects Ever/never[b]			Substantial/never[c]		
	N	OR	95% CI	N	OR	95% CI	N	OR	95% CI	N	OR	95% CI
Oesophagus	23	0.9	0.6–1.5	11	1.1	0.5–2.0	15	0.9	0.5–1.7	8	1.2	0.5–2.6
Stomach	83	1.6	1.2–2.2	38	1.7	1.1–2.6	45	1.4	0.9–2.0	19	1.4	0.8–2.4
Colon	76	0.9	0.7–1.2	37	1.1	0.7–1.6	42	0.8	0.6–1.2	20	1.0	0.6–1.6
Rectosigmoid	58	1.0	0.7–1.4	26	1.0	0.6–1.6	31	1.0	0.6–1.5	13	1.0	0.5–1.9
Rectum	49	1.0	0.7–1.5	21	1.0	0.6–1.6	30	1.2	0.7–1.9	13	1.1	0.5–2.1
Liver	7	0.5	0.2–1.2	1	0.2	0.0–1.4	4	0.5	0.2–1.5	0	0.0	0.0–0.0
Pancreas	27	0.8	0.5–1.3	9	0.6	0.3–1.3	18	1.2	0.7–2.2	5	0.8	0.3–2.2
Lung:												
Oat-cell	49	1.2	0.8–1.8	23	1.4	0.9–2.4	35	1.3	0.8–2.0	15	1.5	0.8–2.8
Squamous-cell	13	1.2	0.9–1.6	60	1.5	1.0–2.1	83	1.3	0.9–1.8	43	1.7	1.1–2.7
Adenocarcinoma	37	0.8	0.5–1.2	15	0.8	0.5–1.5	23	0.7	0.4–1.1	7	0.6	0.2–1.2
Other or unknown type	51	1.2	0.8–1.7	31	1.7	1.1–2.7	43	1.7	1.1–2.6	25	2.6	1.5–4.5
Prostate	99	0.8	0.6–1.1	39	0.8	0.5–1.1	73	1.0	0.7–1.3	31	1.0	0.6–1.5
Bladder	119	1.1	0.9–1.4	48	1.1	0.8–1.5	62	0.9	0.7–1.3	25	0.9	0.6–1.4
Kidney	48	1.2	0.8–1.7	18	1.0	0.6–1.7	29	1.2	0.8–2.0	11	1.1	0.6–2.2
Melanoma	21	0.9	0.6–1.5	8	0.8	0.4–1.8	10	1.2	0.5–2.4	5	1.3	0.5–3.4
Lymphoma	47	1.0	0.7–1.4	23	1.0	0.6–1.6	32	1.1	0.7–1.6	14	1.0	0.5–1.8
Hodgkin's	13	1.2	0.6–2.4	3	0.6	0.2–2.3	9	1.2	0.5–2.8	2	0.6	0.1–2.8

[a] N, number of exposed cases; OR, odds ratio; 95% CI, 95% confidence interval
[b] Exposure at any level for any duration
[c] 'Substantially' exposed subjects only – see p. 35 for definition of substantial exposure.

than that due to silica. Nevertheless, in the two smoking strata containing large enough numbers to provide stable odds ratio estimates, the highest odds ratio is in the subgroup with substantial silica exposure. From a statistical viewpoint, the data fit a multiplicative model (chi-square = 9, with six degrees of freedom), though they would undoubtedly also fit other models.

Taking at face value the odds ratio estimates for stomach and lung cancer in Table 3, and the prevalence of exposure to silica, we estimate the attributable percentage risk due to silica exposure to be 8.1% for stomach cancer and 7.4% for lung cancer (except adenocarcinoma).

Table 3. Odds ratio for associations between silica and two cancer types by duration and degree of exposure, French Canadian subjects only[a]

Exposure category[b]	Stomach cancer			Lung cancer (except adenocarcinoma)		
	N	OR	95% CI	N	OR	95% CI
Not exposed	101	1.0		318	1.0	
Any exposure	45	1.4	1.0-1.9	161	1.3	1.0-1.8
Duration:						
< 10 years	22	1.5	0.9-2.5	55	1.2	0.8-1.8
10-20 years	8	0.9	0.4-2.1	31	1.1	0.6-1.8
> 20 years	15	1.1	0.6-2.0	75	1.7	1.1-2.5
Concentration:						
low	28	1.2	0.8-2.0	97	1.3	1.0-1.9
medium	13	1.2	0.6-2.2	48	1.3	0.8-2.0
high	4	1.1	0.4-3.2	16	1.6	0.8-3.3
Frequency:						
< 5%	2	0.6	0.1-2.6	17	1.3	0.7-2.6
5-30%	24	1.4	0.8-2.2	72	1.1	0.8-1.6
> 30%	19	1.1	0.7-2.0	72	1.7	1.1-2.5
Confidence:						
possible	1	0.8	0.1-6.2	6	0.7	0.2-2.1
probable	11	1.2	0.6-2.4	40	1.1	0.7-1.8
certain	33	1.2	0.8-1.9	115	1.5	1.1-2.0

[a] A separate logistic regression model was fitted for each type of cancer and each dimension of exposure. All of the models included the *a priori* confounders: age, smoking, socioeconomic class, blue/white collar job history and asbestos. The stomach cancer models included, in addition, the empirical confounders: father's social class, residential use of surface water during childhood, engine emissions and solvents. The lung cancer model included, in addition to the *a priori* confounders, the empirical confounders: level of education attained and marital status. N, number of exposed cases; OR, odds ratio; 95% CI, 95% confidence interval
[b] See Methods section for explanation of concentration, frequency and confidence dimensions

Discussion

The present investigation had several advantages over many published studies on silica and cancer risk. The study was population-based and thus covered the spectrum of workplaces in which exposure to silica might occur. These findings may therefore be more generalizable in their implications than those of most cohort studies, which may be relevant only to the particular occupation investigated. Cancer attribution was based on histologically confirmed diagnoses rather than on the more error-prone death certificate reports. Exposure attribution was based on an intensive data collection procedure entailing a probing interview with

Table 4. Odds ratios for associations between lung cancer (except adenocarcinoma) and varying degrees of silica exposure, by smoking status, among French Canadian subjects only[a]

Lifetime smoking status	Lifetime silica exposure[b]					
	Not exposed		Not 'substantial'		'Substantial'	
	N	OR	N	OR	N	OR
Non-smoker	3	1.0	1	2.0	1	2.6
1–59 cig. – years	23	7.1	6	12.5	2	4.5
600–1199 cig. – years	92	14.8	20	12.9	25	28.2
1200+ cig. – years	200	32.7	45	41.9	61	47.5

[a] Odds ratio (OR) computed by logistic regression with model including age, socioeconomic status, overall salubrity of job history, education, marital status, asbestos as well as smoking and silica
[b] Exposure status defined by function of level, frequency, confidence, duration of exposure. See Methods for details.

the patient himself to obtain a detailed complete lifetime job history, followed by an analysis of the job history by chemists and hygienists with a view to determining the likelihood and degree of exposure to silica and other substances. Several factors were included as potential confounders in the screening analyses. In in-depth analyses, several hundred covariates were evaluated as potential confounders and those showing such effects were included in the regression model.

The main limiting factor in this study was the validity of the retrospective exposure assessment. The assessment was based on the judgement of a specially trained and well documented team of technical experts. While components of this process have been evaluated and found to be satisfactory (Baumgarten et al., 1982; Goldberg et al., 1986; Siemiatycki, 1984), no ultimate validation is possible because no relevant hygiene measurements were made in the past in the vast majority of work sites. Whatever misclassification occurred would have had the following effects: if silica exposure was seriously miscoded, it would lead to an attenuation of odds ratio estimates; however, if an important confounding factor were seriously miscoded it would be inadequately adjusted and an exaggerated odds ratio estimate could ensue. Although the system was imperfect, we believe that the quality of our exposure information represents a significant improvement over the subject's job title, which has been the conventional exposure variable used in the past in case–control studies (Siemiatycki et al., 1981; Siemiatycki, 1984).

Since we estimated many odds ratios in the screening analyses and in fact estimated each odds ratio under four sets of conditions, it is possible that some were significant by chance. To help tease out the true from false positives, we subjected the two noteworthy associations to in-depth analyses aimed at eliminating the effects of potential confounding factors and establishing whether there was any 'dose–response' relationship.

While the overall odds ratio for stomach cancer–silica is elevated, the pattern of risk by duration and degree of exposure is not what would be expected in a causal relationship. Thus, our evidence for an increased risk of stomach cancer is not strong. Nevertheless, it is noteworthy that there have been previous reports implicating silica in gastrointestinal cancer (IARC, 1987).

The results of the present investigation provide fairly strong evidence of an association between long-term high-level silica exposure and lung cancer, excluding adenocarcinoma. Because the epidemiology of adenocarcinoma of the lung has often been reported to be different from that of other types of lung cancer (Lubin & Blot, 1984), its exceptional status in our findings does not detract from the plausibility of the associations observed with other cell types. Although we distinguished different degrees of exposure to silica, we are unable to attach to them any absolute numerical values.

Cigarette smoking is the most important potential confounder for the silica–lung cancer association, particularly for the non-adenocarcinomas. Within each of four smoking categories, there was some evidence of increased lung cancer risk with increased silica exposure. Because of the small numbers in the non-smoker and low-smoker categories, the evidence of increased risk for these categories was weaker than for the higher smoking ones.

Had any of the other cancer sites been found to be associated with silica exposure, we would have considered it an interesting but implausible hypothesis. The fact that, of all the sites studied, it was lung cancer that showed the strongest evidence of association, and that this is the site that has been linked with silica exposure in several previous reports (Goldsmith et al., 1982; Heppleston, 1985; IARC, 1987) adds considerably to the credibility of the findings.

Exposure to silica is one of the most prevalent of occupational exposures. Our estimates of the population attributable risk percentage are very high. If as many as 7% of non-adenocarcinoma lung cancers are attributable to silica exposure, this would consitute a major public health problem, as important as asbestos-related lung cancer.

The present report overlaps somewhat with an analysis of the associations between ten inorganic dusts, including silica, and several types of cancer (Siemiatycki et al., 1989). In that report we showed that not only silica but also several other types of inorganic dusts - namely, excavation dust, concrete dust, cement dust, lime dust, alumina - were associated with non-adenocarcinoma of the lung. While there may have been some mutual confounding in those results, they suggest the possibility of a general effect of respirable inorganic dusts as lung carcinogens.

Acknowledgements

This research was supported by grants from the Institut de Recherche en Santé et Sécurité du Travail du Québec, the National Health Research and Development Program, and the National Cancer Institute of Canada. Chemical coding of jobs was carried out by Howard Kemper and Louise Nadon as well as Ramzan Lakhani and Denis Begin, case ascertainment and interviewing by Denise Bourbonnais, Yves Céré, Lucy Felicissimo, Hélène Sheppard, Vincent Varacelli and Michel Vinet. Jean Pellerin and Lorne Wald helped with data management and programming of analysis. This study would not have been possible without the cooperation of the following clinicians and pathologists, and their respective hospital authorities: Drs Y. Méthot, R. Vauclair and Y. Ayoub, Hôpital Notre-Dame; Dr B. Case, Royal Victoria Hospital; Drs C. Lachance and H. Frank, Sir Mortimer B. Davis Jewish General Hospital; Drs W.P. Duguid and J. MacFarlane, Montreal General Hospital; Drs S. Tange and D. Munro, Montreal Chest Hospital; Drs F. Gomes and F. Wiegand, Queen Elizabeth Hospital; Drs B. Artenian and G. Pearl, Reddy Memorial Hospital; Drs D. Kahn and C. Pick, St. Mary's Hospital; Dr C. Piché, Hôpital Ste-Jeanne d'Arc; Drs P. Bluteau and G. Arjane, Centre Hospitalier de Verdun; Drs Yves McKay and A. Bachand, Hôpital du Sacré-Coeur; Drs A. Neaga and A. Reeves, Hôpital Jean-Talon; Drs Y. Boivin and M. Cadotte, Hôtel-Dieu de Montréal; Dr A. Iorizzo, Hôpital Santa Cabrini; Dr A. Bonin, Hôpital Fleury; Drs J. Lamarche and G. Lachance, Hôpital Maisonneuve-Rosemont; Drs G. Gariépy and S. Legault-Poisson, Hôpital St-Luc; Dr M. Mandavia, Lakeshore General Hospital; Dr J.C. Larose, Cité de la Santé. The authors thank the pathology departments and tumour registry staff of the above-mentioned hospitals who notified us of incident cases.

References

Baker, R.J. & Nelder, J.A. (1978) *The GLIM System. Release 3. Generalised linear interactive modelling*, Oxford, Royal Statistical Society

Baumgarten, M., Siemiatycki, J. & Gibbs, G.W. (1983) Validity of work histories obtained by interview for epidemiologic purposes. *Am. J. Epidemiol.*, 118, 583-591

Bruzzi, P., Green, S.B., Byar, D.P., Brinton, L.D. & Schairer, C. (1985) Estimating the population attributable risk for multiple risk factors using case-control data. *Am. J. Epidemiol.*, 122, 904-914

Dewar, R.A.D. & Siemiatycki, J. (1985) A program for point and interval calculation of odds ratios and attributable risks from unmatched case-control data. *Int. J. Biomed. Comput.*, 16, 183-190

Gérin, M., Siemiatycki, J., Kemper, H. & Begin, D. (1985) Obtaining occupational exposure histories in epidemiologic case-control studies. *J. Occup. Med.*, 27, 420-426

Goldberg, M., Siemiatycki, J. & Gérin, M. (1986) Inter-rater agreement in assessing occupational exposure in a case-control study. *Br. J. Ind. Med.*, 43, 667-676

Goldsmith, D.F., Guidotti, T.L. & Johnston, D.R. (1982) Does occupational exposure to silica cause lung cancer? *Am. J. Ind. Med.*, 3, 423-440

Heppleston, A.G. (1985) Silica, pneumoconiosis, and carcinoma of the lung. *Am. J. Ind. Med.*, 7, 285-294

IARC (1987) *IARC Monographs on the Evaluation of the Carcinogenic Risk of Chemicals to Humans*, Vol. 42, *Silica and Some Silicates*, Lyon, International Agency for Research on Cancer

Kurppa, K., Koskela, R.S. & Gudbergsson, H. (1982) Gastrointestinal cancer in workers exposed to quartz. *Lancet*, ii, 150

Lubin, J.H. & Blot, W.J. (1984) Assessment of lung cancer risk factors by histologic category. *J. Natl Cancer Inst.*, 73, 383-389

Mantel, N. & Haenszel, W. (1959) Statistical aspects of the analysis of data from retrospective studies of disease. *J. Natl Cancer Inst.*, 22, 719-748

Ministry of Manpower and Immigration (1974) *Canadian Classification and Dictionary of Occupations - 1971*, Ottawa, Information Canada

Siemiatycki, J. (1984) An epidemiologic approach to discovering occupational carcinogens by obtaining better information on occupational exposures. *Recent Adv. Occup. Health*, 2, 143-157

Siemiatycki, J., Day, N. Fabry, J. & Cooper, J.A. (1981) Discovering carcinogens in the occupational environment: a novel epidemiologic approach. *J. Natl Cancer Inst.*, 66, 217-225

Siemiatycki, J., Gérin, M., Richardson, L., Hubert, J. & Kemper, H. (1982) Preliminary report of an exposure-based, case-control monitoring system for discovering occupational carcinogens. *Teratog. Carcinog. Mutag.*, 2, 169-177

Siemiatycki, J., Wacholder, S., Richardson, L., Dewar, R. & Gérin, M. (1987) Discovering carcinogens in the occupational environment: methods of data collection and analysis of a large case-referent monitoring system. *Scand. J. Work Environ. Health*, 13, 486-492

Siemiatycki, J., Dewar, R., Lakhani, R., Nadon, L., Richardson, L. & Gérin, M. (1989) Cancer risks associated with ten inorganic dusts: Results from a case-control study in Montreal. *Am. J. Ind. Med.* (in press)

Cancer mortality of granite workers 1940-1985

R.-S. Koskela, M. Klockars, E. Järvinen, A. Rossi and P.J. Kolari

Institute of Occupational Health, Helsinki, Finland

Summary. A retrospective cohort study was undertaken to investigate the cancer mortality of granite workers. The study comprised 1026 workers who took up such work between 1940 and 1971. The number of person-years was 23 434, and the number of deaths was 296. During the total follow-up period, 59 tumours were observed as compared with 54.4 expected. An excess mortality from tumours was observed in workers followed up for 20 years or more. Of the 59 tumours, 31 were lung cancers (expected 19.9), and 18 gastrointestinal cancers (expected 11.6), nine of which were stomach cancers (expected 7.1). Mortality from lung cancer was excessive for workers followed up for at least 15 years (28 observed, 12.7 expected). The results indicate that granite exposure *per se* may be an etiological factor in the initiation or promotion of malignant neoplasms.

Introduction

Epidemiological data have suggested an excess of lung cancer mortality among silicotic patients (Westerholm, 1980; Gudbergsson *et al.*, 1984; Schüler & Rüttner, 1986) but the evidence for a causal relationship between pure silica exposure and cancer is inconclusive (Davis *et al.*, 1983; Steenland & Beaumont, 1986). The International Agency for Research on Cancer concluded in June 1986 that there is 'limited evidence' for the carcinogenicity of crystalline silica to humans (IARC, 1987). The efforts of epidemiologists to determine whether a cause-effect relationship exists between silica exposure, silicosis and lung cancer have been hampered by differences in the extent of confounding exposures, e.g., to radon daughters in mining and to polycyclic aromatic hydrocarbons in foundries (Axelson & Sundell, 1978; Katsnelson & Mokronosova, 1979; Fox *et al.*, 1981; Goldsmith *et al.*, 1982) as well as by the diversity of epidemiological study designs. Variability in study design as between, e.g., cohort and case-referent studies, or the inclusion either of workers exposed to silica or only those with manifest silicosis in the study population makes it extremely difficult to compare the results obtained in different studies.

The present paper summarizes the results of our follow-up study on the cancer mortality of granite workers, a cohort which has been exposed to 'pure' silica, without significant exposure to radon, asbestos, polycyclic aromatic hydrocarbons, etc., and has been followed up for at least 14 years (the last year of entry was 1971).

Materials and methods

Subjects

The cohort comprised 1026 Finnish granite workers employed in quarries and processing yards from three regions of Finland, who entered such work between 1940 and 1971; all were employed for at least three months.

The workers' names, dates and places of birth, and work history at each firm were collected from the employers' personnel records, and all were traced through the Population Data Register. Up to the end of 1985, a total of 296 had died. The occupational status of the granite workers on 31 December 1985 is shown in Figure 1. The causes of death were ascertained from the death certificates and were coded according to the Eighth Revision of the *International Classification of Diseases*.

Figure 1. Occupational status of granite workers on 31 December 1985

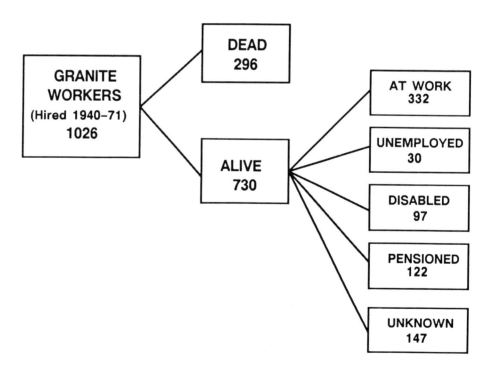

The mortality of the cohort had earlier been followed up until the end of 1972 (Koskela & Järvinen, 1975), 1975 (Kurppa et al., 1982) and 1981 (Koskela et al., 1987a,b). In the present study, in which the cohort was followed up until the end of 1985, the official statistics of causes of death for 1975, the median year for deaths in the cohort during 1940–1985, were compared with those for the general male population (World Health Organization, 1977). The median year was chosen because of the similarity between the mortality figures calculated in our earlier analyses for the median year for deaths and five-year calendar period-specific mortality rates, expressed both as standardized mortality ratios (SMRs) and proportional mortality rates (PMRs) (Koskela et al., 1984; Tola et al., 1979). We also ascertained the stability of the lung cancer mortality rates in the present study by calculating the SMRs based on the mortality of the general population for a number of years (1964, 1970, 1980 and 1985). In addition, we calculated the PMRs both for the median year for deaths (1975) and separately for ten-year calendar periods during the follow-up (1940–1985). The differences between the results of the two calculations were negligible. The use of the median year even slightly overestimated the expected numbers.

A questionnaire covering both the complete work history and smoking habits was sent either to the workers themselves, if they were still alive, or to the next of kin if they were not, the response rate being 73% in both cases. Work and smoking histories were also obtained for the non-respondents either from the 1972 questionnaire or from medical records. It was therefore possible to obtain more complete work histories for about 90% of the entire cohort.

Exposures

The mean duration of exposure to quartz dust was about 12 years. Dust exposure in the Finnish granite industry was investigated during the period 1970–1972 (Ahlman et al., 1975). Dust concentrations were found to be high in several stages of the process. The geometric mean of the total dust concentration ranged from 1.7 to 39.8 mg/m^3 and of quartz dust from 1.0 to 1.5 mg/m^3. The highest concentrations were found in drilling, where the hygiene standard for quartz was exceeded, on the average, by a factor of ten. The job title of each granite worker was obtained from the employers' personnel records. The duration of exposure to granite dust was assessed from the replies to the questionnaire.

Potential confounders

The geographical location of the workplaces and the fact that the work is done out of doors probably excludes exposure to major confounders such as radiation, asbestos or polycyclic aromatic hydrocarbons produced during combustion (United Nations, 1982). Confounding occupational exposure either before or after employment as granite workers can also be considered as unimportant, based on the replies to the questionnaire.

Employment in occupations that might involve exposure to possible concomitant carcinogens was reported by 18% of the total cohort (Table 1). The mean confounding exposure time was two years. We checked, for each individual, from the answers to the questionnaire, the quality and duration of potential confounding exposure. Of the 181 workers with potential confounding exposure, 120 (11%

Table 1. Occupational exposure of granite workers to possible carcinogens other than silica[a]

Occupational exposure	No.	%
Non-confounding exposure	630	61
Possible confounding exposure:	181	18
Mining	6	0.6
Foundry	29	2.8
Insulation (rock-wool)	1	0.1
Glass-work	1	0.1
Brick-making	4	0.1
Other dusty work	37	3.6
Welding	49	4.8
Road construction	5	0.5
Other gases[b]	8	0.8
Chemicals (e.g., solvents)	26	2.5
Multiple exposures	15	2.5
Unknown	215	21
Total cohort	1026	100

[a] Based on complete occupational histories obtained in 1985
[b] Excludes welding and asphalt work

of the cohort) proved to have such an exposure. Only 33 workers (3% of the cohort) had obvious confounding exposure, i.e., at least five years' exposure to some other carcinogen and less than five years' exposure to granite dust. We have determined the maximum effect of possible confounding (concomitant carcinogenic) exposure by calculating the mortality rates after all the workers who had reported such exposure had been excluded (see Table 1).

Data on smoking habits were collected by means of the questionnaire in the course of the health screening of current employees in 1970–1972. Non-smokers made up 21%, ex-smokers 17% and smokers 62%. At the end of 1985, 15% were non-smokers, 60% had smoked at some time, while for 25% no data on smoking were available (Table 2). The smoking habits of the granite workers are thus similar to those of other groups of male Finnish workers of the same age (Asp, 1984; Martelin, 1984).

Statistical analysis

In an analysis of the mortality pattern, age-specific observed and expected numbers of deaths based on national figures and SMRs were computed for the different causes of death. The observed and expected numbers were subdivided into five-year follow-up periods and calculated according to years since entry (latency period). The observed age-specific and cause-specific numbers of deaths were compared by means of the Poisson distribution model with the corresponding expected numbers (Lentner, 1982).

Table 2. Smoking history of granite workers

Smoking category	Alive		Deceased	
	No.	%	No.	%
Non-smokers	123	23	28	14
Smokers:	414	77	181	86
Moderate smokers				
(< 20 cigarettes per day)	65	12	13	6
Heavy smokers				
(> 20 cigarettes per day)	349	65	168	80
Data obtained	537	100	209	100
Data missing	193		87	
Total	730		296	

Results

By the end of 1985, 296 of the 1026 granite workers had died. The observed and expected numbers of deaths for the total follow-up period (1940–1985) are shown in Table 3. Both the overall mortality and cause-specific mortality were slightly higher at the end of 1985 than at the end of previous follow-up periods (Table 4). The observed number of tumours was 59 and the expected 54.4 (Table 3). Mortality from cancer of the digestive organs was 1.6 times the expected value (observed 18, expected 11.6). Nine of these cases were cancer of the stomach, the corresponding expected number being 7.1. The other gastrointestinal cancers were cancer of the oesophagus (three cases), cancer of the biliary tract (one case), cancer of the pancreas (three cases), and non-defined cancer (two cases). Tumours other than those of the gastrointestinal system and lungs, of which there were ten, were not predominantly of any particular type.

Table 3. Observed and expected numbers of deaths, 1940–1985[a]

Cause of death	Observed	Expected	SMR
Cardiovascular diseases	128	135.0	95
Tumours:	59	54.4	108
Lung cancer	31	19.9	156[b]
Cancer of digestive organs	18	11.6	155
Respiratory diseases	40	17.8	255[c]
Total	296	276.2	107

[a] Period of entry 1940–71
[b] $p < 0.05$
[c] $p < 0.001$, Poisson distribution

There were 31 deaths due to lung cancer (expected 19.9). Of the 40 deaths due to respiratory diseases, 13 were due to silicosis. From 1940 to 1985, only three decedents had had both lung cancer and silicosis (in two cases silicosis was

Table 4. Standardized mortality ratios for the cohort of granite workers at the end of the follow-up period[a]

Cause of death[b]	1972	1975	1981	1985
Cardiovascular diseases	71[c]	70[c]	87	95
Tumours:	86	100	102	108
Lung cancer	44[d]	74	129	156[c]
Cancer of digestive organs	185	203[c]	155	155
Respiratory diseases	200[c]	265[e]	201[e]	255[e]
Total	91	94	102	107

[a] Period of entry 1940–71
[b] A detailed mortality analysis of the cohort has been given by Koskela et al., 1987a
[c] $p < 0.05$
[d] Based on only four subjects
[e] $p < 0.001$, Poisson distribution

the secondary cause). Lung cancer mortality calculated for separate follow-up periods showed an excess after 15 years since entry into granite work. This is also seen when the data are analysed by latency period (Table 5). The mean exposure time for lung cancer cases was 19 years.

Table 5. Observed and expected lung cancer deaths by years since entry into granite work (latency)

Years since entry (latency)	Lung cancer deaths		
	Observed	Expected	SMR
0	31	19.9	156[a]
≥ 5	31	18.4	168[a]
≥ 10	31	16.1	193[b]
≥ 15	28	12.7	220[c]
≥ 20	20	8.9	225[b]
≥ 25	16	5.7	281[c]
≥ 30	7	3.1	226

[a] $p < 0.05$
[b] $p < 0.01$
[c] $p < 0.001$, Poisson distribution

Discussion

The cohort in the present study comprised outdoor granite workers exposed to relatively pure silica. They were not exposed to other carcinogens that might affect the outcome. Of the 59 cancer cases, five had potential confounding exposure either before or after employment as granite workers, namely, three of the lung cancer cases (of which one was obvious), none among the gastrointestinal cancer cases, and two among other cancer cases (both of them obvious). There was no appreciable change in the cancer risk when workers with concomitant carcinogenic exposure were excluded. The SMR for lung cancer for the total cohort was originally 156; after exclusion of confounding exposure (see Table 1) it was 153.

The smoking habits of granite workers were not markedly different from those of male Finnish industrial workers in general. Of the 31 cases with lung cancer, 29 had been smokers, while for two cases no data were available. The observed rate ratios for lung cancer during separate follow-up periods ranging from 15 years to more than 35 varied between 1.2 and 3.8. These ratios do not support the possibility that smoking alone can explain the excessive risk of lung cancer because the maximum confounding effect of smoking would not, even theoretically, be more than about two-fold as compared with the general population (Axelson, 1978). The analysis of the complete work histories showed that 17% of the granite workers had entered the industry before 1940, the criterion year for first entry into the cohort (Figure 2).

Figure 2. Annual cohort entries of granite workers based on employers' records and questionnaires

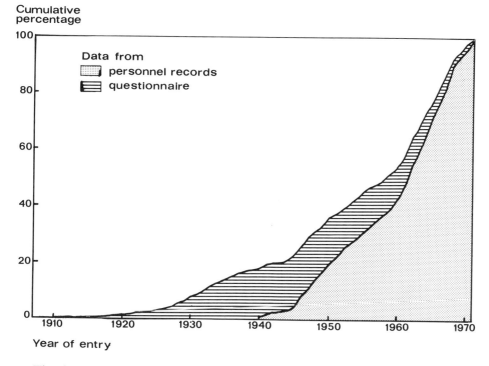

The duration of exposure ranged from 0 to 49 years (mean 12 years). According to the employers' personnel records, the range of exposure duration was 0–31 years and the mean four years. We compared person-years calculated from complete work histories with those based on the personnel records of the employers. The coverage of person-years in the employers' records was lowest (65–75%) for the young age groups (< 35 years) and highest (87–100%) for the older groups (≥ 45 years), as compared with those reported by the employees. Because of the good coverage of person-years among the oldest workers, the effect of loss of person-years on cancer mortality rates was small.

Because the complete work histories also made it necessary to make changes in the calculation of latency periods, we relocated the deaths in different follow-up periods on the basis of the questionnaire data. Consequently, the deaths were transferred to later follow-up periods than those in which they had originally been located. About half the workers who had died from cancer had entered granite work before 1940. When the mortality rates, e.g., for lung cancer, were calculated on the basis of complete work histories (both person-years and deaths), they were lower than those based on employers' records, but remained statistically significant. We also ascertained the stability of our mortality results by excluding from the cohort all persons who had entered granite work before 1940 (according to the complete work history). The SMR for lung cancer for the total cohort was then practically the same (151) as the original SMR (156).

Excess mortality from lung cancer has been reported in a cohort study of silicotic stone workers exposed to pure silica (Gudbergsson et al., 1984). Steenland and Beaumont (1986) also found an excess risk of lung cancer among a subgroup of silicotic patients who had been exposed to pure silica. A case–referent study of silicotic patients showed an increased lung cancer risk for smokers, but not for non-smokers (Zambon et al., 1983). Whether the association found in these studies is due to the effect of quartz dust itself or to the fibrotic silicosis process is unclear. The interpretation of the results has often been complicated by methodological differences between the different studies, e.g., differences in study design, the internal structure of the cohort, potential confounders, methods of analysis, and exposure intensity and duration (Koskela et al., 1984; Hoel, 1985).

In the present study, the periods of exposure and follow-up were sufficiently long for both silicosis and cancer to develop. Manifest pulmonary silicosis was not very common. In fact, it is remarkable that only three patients with cancer also had pulmonary silicosis. The increased mortality previously observed from cancer of extrapulmonary sites (gastrointestinal cancer) (Table 4), certain non-malignant disease patterns (Koskela et al., 1987a; Klockars et al., 1987), as well as the complete work histories of the cancer patients support the hypothesis of a direct association between silica exposure and lung cancer.

A simplified model of the consequences of quartz exposure *in vivo* is shown in Figure 3. The proposed mechanisms include the following: (1) silica exposure directly induces lung cancer ($X \rightarrow Y$); (2) silica causes chronic silicosis, which may be an intermediate pathological state leading to lung cancer ($X \rightarrow W \rightarrow Y$); and (3) silica combined with potential carcinogens ($X \rightarrow Z \rightarrow Y$, where Z = cigarette smoke, radon, polycyclic aromatic hydrocarbons, etc.) leads to the development of neoplasia (Goldsmith & Guidotti, 1986). However, it is extremely difficult to take into account all known or possible environmental factors. The theoretical pathways whereby pulmonary neoplasia can arise are numerous. Carcinogenesis is a multistage process, and potential molecular targets for quartz include DNA damage and repair, effects of altered gene products, activation of growth factors, the stimulation of macrophage-derived growth regulatory factors, etc. (for review, see Saffiotti, 1986). In experimental animals, the fibrogenic reaction is accompanied by hyperplasia of the epithelial lining, a phenomenon not observed in areas of the lung not affected by fibrosis. It has been suggested that the induction of preneoplasia associated with fibrosis, e.g., the development of epithelial dysplasia adjacent to fibrotic areas, is the result of the structural disorder brought about by fibrosis in both intracellular communication and homeostatic growth regulation (Brand, 1986).

Figure 3. A causal model of lung cancer

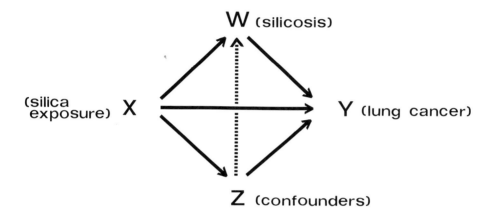

Biological interaction between occupational silica exposure and other promoting or confounding factors and exposures is likely. For instance, the 'confounded' pathway (Figure 3) may involve the adsorption of carcinogenic substances on to the surface of silica, and thence to the deposition of these substances in the lung. In addition, inhaled chemical carcinogens deposited in the lung may not be removed by normal clearing mechanisms if these have been adversely affected by even slight local fibrotic lesions. This might result in a synergistic effect of silica and smoking in the development of lung cancer (Goldsmith & Guidotti, 1986).

However, the hypothesis that silica is itself a carcinogen is supported by observations showing that it causes both chromosomal aberrations and the neoplastic transformation of cells *in vitro* (Hesterberg et al., 1986), and by epidemiological studies (Goldsmith et al., 1982), as well as by the results of our study on granite workers.

Acknowledgements

We thank the Central Statistical Office of Finland for the data on causes of deaths, and the Finnish Cancer Registry for the cancer mortality data.

References

Ahlman, K., Backman, A.L., Hannunkari, I., Järvinen, E., Koponen, E., Koskela, R.S., Partanen, T., Seppäläinen, A.M. & Starck, J. (1975) *Work conditions and health of granite workers* [in Finnish with English summary], Helsinki, Kansaneläkelaitoksen julkaisuja

Asp, S. (1984) Confounding by variable smoking habits in different occupational groups. *Scand. J. Work Environ. Health*, *10*, 325-326

Axelson, O. (1978) Aspects of confounding in occupational health epidemiology. *Scand. J. Work Environ. Health*, *4*, 85-89

Axelson, O. & Sundell, L. (1978) Mining, lung cancer and smoking. *Scand. J. Work Environ. Health*, *4*, 46-52

Brand, K.G. (1986) Fibrotic scar cancer in the light of foreign body tumorigenesis. In: Goldsmith, D.F., Winn, D.M. & Shy, C.M., eds, *Silica, Silicosis and Cancer: Controversy in Occupational Medicine* (Cancer Research Monographs, Vol. 2), New York, NY, Praeger, pp. 281-286

Davis, L.K., Wegman, D.H., Monson, R.R. & Froines, J. (1983) Mortality experience of Vermont granite workers. *Am. J. Ind. Med.*, *4*, 705-723

Fox, A.J., Goldblatt, P. & Kinlen, L.J. (1981) A study of the mortality of Cornish tin miners. *Br. J. Ind. Med.*, *38*, 378-380

Goldsmith, D.F., Guidotti, T.L. & Johnston, D.R. (1982) Does occupational exposure to silica cause lung cancer? *Am. J. Ind. Med.*, *3*, 423-440

Goldsmith, D.F. & Guidotti, T.L. (1986) Combined silica exposure and cigarette smoking: a likely synergistic effect. In: Goldsmith, D.F., Winn, D.M. & Shy, C.M., eds, *Silica, Silicosis and Lung Cancer: Controversy in Occupational Medicine* (Cancer Research Monographs, Vol. 2), New York, NY, Praeger, pp. 451-459

Gudbergsson, H., Kurppa, K., Koskinen, H. & Vasama, M. (1984) An association between silicosis and lung cancer. A register approach. In: *Proceedings of the VIth International Pneumoconiosis Conference 1983, Bochum, Federal Republic of Germany, 20-23 September*, Vol. 1, Bochum, Bergbau-Berufsgenossenschaft, pp. 212-216

Hesterberg, T.W., Oshimura, M., Brody, A.R. & Barrett, J.C. (1986) Asbestos and silica induce morphological transformation of mammalian cells in culture: a possible mechanism. In: Goldsmith, D.F., Winn, D.M. & Shy, C.M., eds, *Silica, Silicosis and Lung Cancer: Controversy in Occupational Medicine* (Cancer Research Monographs, Vol. 2), New York, NY, Praeger, pp. 177-190

Hoel, D.G. (1985) The impact of occupational exposure patterns on quantitative risk estimation. In: *Risk Quantification and Regulatory Policy*, Cold Spring Harbor, NY, Cold Spring Harbor Laboratory (Banbury Report 19), pp. 105-118

IARC (1987) *IARC Monographs on the Evaluation of the Carcinogenic Risk of Chemicals to Humans:* Vol. 42, *Silica and Some Silicates*, Lyon, International Agency for Research on Cancer

Katsnelson, B.A. & Mokronosova, K.A. (1979) Non-fibrous mineral dusts and malignant tumours: an epidemiological study of mortality. *J. Occup. Med.*, *21*, 15-20

Klockars, M., Koskela, R.S., Järvinen, E., Kolari, P.J. & Rossi, A. (1978) Silica exposure and rheumatoid arthritis: a follow-up study of granite workers 1940-81. *Br. Med. J.*, *294*, 997-1000

Koskela, R.S. & Järvinen, E. (1975) Mortality, disability and health among granite workers. In: Ahlman, K., Backman, A.L., Hannunkari, I., Järvinen, E., Koponen, M., Koskela, R.S., Partanen. T., Seppäläinen, A.M. & Starck, J. *Work conditions and health of granite workers* [in Finnish], Helsinki, Kansaneläkelaitoksen julkaisuja

Koskela, R.-S., Järvinen, E. & Kolari, P.J. (1984) Effect of cohort definition and follow-up length on occupational mortality rates. *Scand. J. Work Environ. Health*, *10*, 311-316

Koskela, R.S., Klockars, M., Järvinen, E., Kolari, P.J. & Rossi, A. (1987a) Mortality and disability among granite workers. *Scand. J. Work Environ. Health*, *13*, 18-25

Koskela, R.S., Klockars, M., Järvinen, E., Kolari, P.J. & Rossi, A. (1987b) Cancer mortality of granite workers. *Scand. J. Work Environ. Health*, *13*, 26-31

Kurppa, K., Koskela, R.S. & Gudbergsson, H. (1982) Gastrointestinal cancer in workers exposed to quartz. *Lancet*, *i*, 150

Lentner, C., ed., (1982) *Geigy Scientific Tables*, Vol. 2, 8th ed., Basel, Ciba-Geigy Limited

Martelin, T. (1984) *The development of smoking habits according to survey data in Finland* [in Finnish], Helsinki, National Board of Health

Saffiotti, U. (1986) The pathology induced by silica in relation to fibrogenesis and carcinogenesis. In: Goldsmith, D.F., Winn, D.M. & Shy, C.M., eds, *Silica, Silicosis and Lung Cancer: Controversy in Occupational Medicine* (Cancer Research Monographs, Vol. 2), New York, NY, Praeger, pp. 287-307

Schüler, G. & Rüttner, J. (1986) Silicosis and lung cancer in Switzerland. In: Goldsmith, D.F., Winn, D.M. & Shy, C.M., eds, *Silica, Silicosis and Lung Cancer: Controversy in Occupational Medicine* (Cancer Research Monographs, Vol. 2), New York, NY, Praeger, pp. 357-366

Steenland, K. & Beaumont, J. (1986) A proportionate mortality study of granite workers. *Am. J. Ind. Med.*, 9, 189-201

Tola, S., Koskela, R.-S., Hernberg, S. & Järvinen, E. (1979) Lung cancer mortality among iron foundry workers. *J. Occup. Med.*, 21, 753-760

United Nations (1982) *Ionizing Radiation: Sources and Biological Effects*, New York, NY, United Nations Scientific Committee on the Effects of Atomic Radiation

Westerholm, P. (1980) Silicosis. Observations on a case register. *Scand. J. Work Environ. Health*, 6, Suppl. 2, 1-86

World Health Organization (1977) *World Health Statistics Annual 1975*, Vol. 1, *Vital Statistics and Causes of Death*, Geneva

Zambon, P., Mastrangelo, G. & Saia, B. (1983) Exposure to silica and lung cancer, a case-control study (paper presented at the Montreal Conference on Occupational Health, 23-25 August 1983)

A mortality study of a cohort of slate quarry workers in the German Democratic Republic

W.H. Mehnert[1], W. Staneczek[1], M. Möhner[1], G. Konetzke[2], W. Müller[3], W. Ahlendorf[3], B. Beck[2], R. Winkelmann[4] and L. Simonato[4]

[1]*National Cancer Registry of the Central Institute for Cancer Research of the Academy of Sciences of the German Democratic Republic, Berlin*
[2]*Central Institute of Occupational Medicine of the German Democratic Republic, Berlin*
[3]*Inspectorate of Industrial Hygiene, Gera County Council, Gera, German Democratic Republic*
[4]*International Agency for Research on Cancer, Lyon, France*

Summary. Following reports from other countries indicating an excess risk of lung cancer among silicotics, a historical cohort of workers employed for at least one year at a company in charge of slate extraction and processing during the period 1953–1985 in the German Democratic Republic has been constructed and followed up for mortality from 1970 to 1985. The results of the study show a mortality excess for infectious and respiratory diseases. The overall lung cancer mortality is not in excess but shows a tendency to increase with time since first exposure. A mortality excess from lung cancer is concentrated among workers receiving compensation for silicosis, suggesting a possible carcinogenic risk for individuals suffering from this pathological condition.

Introduction

The association between cancer, in particular of the lung, and exposure to silica dust, an association which in the past has generally been considered not to exist (Rüttner & Heer, 1969), has in recent years been once again drawn to the attention of cancer researchers. The revival of interest in this field is the consequence of the results of certain epidemiological studies on silicotics which tend to show an excess risk for lung cancer (Finkelstein *et al.*, 1982). The interpretation of these findings in terms of causality is complicated, however, as both the mechanism of selection and the competing risk for other causes of death are probably involved. In view of the large size of the working populations exposed to silica dust and the recent suggestions of a possible cancer risk, studies in this field, aimed at investigating the possible carcinogenic hazard of occupational exposure to silica dust, would seem to be justified.

Following reports from other countries indicating an excess risk of lung cancer among silicotics, a historical cohort of workers employed for at least one year at a company in charge of slate extraction and processing during the period 1953-85 in the German Democratic Republic has been constructed and followed up for mortality from 1970 to 1985.

Materials and methods

The slate quarries include nine different workplaces in three counties in two districts and were combined to form a single company during the early years of the German Democratic Republic. The data for the workplaces were complete and available from 1953 onwards.

The quarries are located in the south of the country (near the border with Czechoslovakia and the Federal Republic of Germany); slate quarries have existed for centuries in this area.

The study population consists of all males employed for at least one year between 1 January 1953 and 31 December 1985 in any of the nine quarries. The observation period extends from 1 January 1970 to 31 December 1985.

The information obtained from the company was completed by the data obtained from the registry of silica-exposed workers and silicotics, which has been in operation since 1950. A complete occupational history for the working period involving exposure to silica dust was available for each identified individual.

Detailed job codes (over 80) were available from the company files; these were subsequently grouped together to form a few main occupational categories. The data were then reviewed by a specialist in dust control with long experience in this field, who classified each worker's exposure to silica dust as high or low. The classification was based mainly on the occupational categories, the data available from airborne dust measurements, and the duration of employment.

Vital status was ascertained from a number of different sources, namely the company records for those still employed at death or alive, the association of former employees for those retired, the silicosis register for those receiving compensation for silicosis, and the police for those who could not be otherwise traced.

As death certificates in the German Democratic Republic are kept only for 15 years, it was possible to establish the causes of death only of those dying between 1 January 1970 and 31 December 1985. All members of the cohort but eight were successfully traced, while the cause of death was established for all individuals known to be dead.

Age- and sex-specific national mortality rates have been used for computing the expected numbers of deaths. A computer program (person-years, version 1.2, February 1986) (Coleman et al., 1986) has been used for computing expected numbers of deaths and standardized mortality ratios (SMRs) for the period of observation 1970-85. The 95% confidence limits of the SMRs have been calculated on the assumption of a Poisson distribution.

Results

The vital status and distribution of the members of the cohort together with their contribution to person-years by time since first exposure are presented in Tables 1 and 2. It is noticeable that almost half (45%) of the total person-years are accumulated after 20 years since first exposure due to the definition of the period at risk. The mortality of the cohort by detailed causes is presented in Table 3. Mortality from diseases of the respiratory system is in excess due to deaths from silicosis. The same applies to the mortality from infectious diseases, which includes deaths from silicosis.

Table 1. Study population by vital status

Vital status	No. of subjects	%
Alive at end of follow-up	2088	84.1
Dead	387	15.6
Lost to follow-up	8	0.3
Total	2483	100

Table 2. Number of subjects and person-years at risk by time since first exposure (based on a total of 2483 subjects)

Item	Time since first exposure (years)				Total
	0-9	10-19	20-29	30+	
No. of subjects	1 236	1 697	1 456	746	2 483
Person-years	7 049	10 032	10 229	3 754	31 064

The mortality from lung cancer is close to the expected figure, while significantly elevated SMRs are reported for rectal cancer and for other neoplasms of the lymphatic and haematopoietic tissue.

When some selected causes of death are analysed by time since first exposure (Table 4), an increase is seen for respiratory diseases, lung cancer and rectal cancer, while mortality from stomach cancer tends to decrease.

The mortality of those members of the cohort who survived for at least 20 years after first exposure is analysed by duration of employment in Table 5 for some selected causes of death. The mortality from both lung cancer and from respiratory diseases tends to increase with duration of employment. The same trend is discernible in Table 6, where the mortality is analysed by qualitative levels of exposure. Tables 7 and 8 present the mortality by detailed causes in silicotics and in non-silicotics respectively.

It is clear that, both overall and for specific causes of death, with the exception of rectal cancer, workers compensated for silicosis have a higher mortality. In

Table 3. Mortality by detailed cause[a]

Cause of death[b]	O	E	SMR	95% CI
All causes (000-999)	387	382.58	101	(91-112)
Infectious diseases (000-136):	7	2.72	258	(104-531)
Tuberculosis of the respiratory system (010-012)	5	1.33	376	(122-877)
Other infectious diseases	2	1.39	144	(17-521)
Malignant neoplasms (140-207):	77	76.64	100	(79-126)
Buccal cavity and pharynx (140-149)	3	1.46	205	(42-600)
Oesophagus (150)	0	1.51	0	(0-245)
Stomach (151)	13	11.19	116	(62-199)
Intestine, except rectum (152-153)	3	3.77	80	(16-233)
Rectum (154)	12	4.56	263	(136-460)
Larynx (161)	0	1.05	0	(0-352)
Trachea, bronchus and lung (162)	27	24.71	109	(72-159)
Leukaemia (204-207)	0	2.28	0	(0-162)
Other neoplasms of the lymphatic and haematopoietic tissue (200-203, 208-209)	8	2.53	316	(136-623)
Other malignant neoplasms	11	23.59	47	(23-83)
Benign and unspecified neoplasms (210-239)	0	1.32	0	(0-279)
Diseases of the circulatory system (390-458)	154	184.36	84	(71-98)
Diseases of the respiratory system (460-519):	74	32.75	226	(177-284)
Pneumoconiosis due to silica and silicates (515)	40	-	-	-
All other known causes	70	84.78	83	(64-104)
Unknown causes	5	-	-	-

[a] O, observed; E, expected; SMR, standardized mortality ratio; 95% CI, 95% confidence interval
[b] Coded in accordance with the Eighth Revision of the *International Classification of Diseases*

particular, the mortality excesses from respiratory diseases and from lung cancer are concentrated among silicotics. When lung cancer mortality is analysed by duration of employment in silicotics and non-silicotics after 20 years since first exposure (Table 9), an increase with time is seen among the former. Among non-silicotics, there is neither an overall excess nor a clear trend, although there is an increase of some 30% among workers with at least 20 years of employment. Similarly, the analysis by qualitative levels of exposure, although limited by small numbers, shows that the excess is concentrated among silicotics, irrespective of the degree of exposure (Table 10).

Table 4. Mortality by time since first exposure[a]

Cause of death	Time since first exposure (years)				Total
	0-9	10-19	20-29	30+	
All causes	13/18.92	48/62.16	166/164.3	160/137.1	387/382.6
	69	77	101	11	101
	(37-117)	(57-102)	(86-118)	(99-136)	(91-112)
All malignant	3/3.37	11/12.96	38/34.19	25/26.13	77/76.64
neoplasms	89	85	111	96	100
	(18-260)	(42-152)	(79-153)	(62-141)	(79-126)
Stomach cancer	1/0.39	3/1.85	7/5.16	2/3.80	13/11.19
	258	163	136	53	116
	(7-1436)	(34-475)	(55-280)	(6-190)	(62-199)
Rectal cancer	0/0.15	1/0.70	5/2.01	6/1.70	12/4.56
	0	143	249	354	263
	(0-2427)	(4-799)	(81-580)	(130-770)	(136-460)
Lung cancer	0/0.87	2/4.01	12/11.28	13/8.54	27/24.71
	0	50	106	152	109
	(0-425)	(6-180)	(55-186)	(81-260)	(72-159)
Lymphomas	1/0.23	3/0.55	4/1.06	0/0.70	8/2.53
	444	546	377	0	316
	(11-2476)	(113-1597)	(103-966)	(0-528)	(136-623)
Respiratory	0/0.97	6/4.67	29/14.60	39/12.52	74/32.75
diseases		129	199	311	226
	(0-382)	(47-280)	(133-285)	(221-426)	(17-284)

[a] The ratio of observed to expected number of deaths, the standardized mortality ratio and (in parentheses) the 95% confidence interval are shown for each cause of death and time since first exposure

Table 5. Mortality by duration of employment of members of the cohort who survived for 20 years after first exposure[a]

Cause of death	Duration of employment (years)			Total
	1-9	10-19	20+	
All causes	54/54.80	114/102.3	158/144.37	326/301.4
	99	111	109	108
	(74-129)	(92-134)	(93-128)	(97-121)
All malignant	8/10.42	21/18.05	34/31.84	63/60.32
neoplasms	77	116	107	104
	(33-151)	(72-178)	(74-149)	(80-134)
Lung cancer	2/3.27	6/5.72	17/10.83	25/19.82
	61	105	157	126
	(7-221)	(39-228)	(91-251)	(82-186)
Respiratory	10/4.48	24/10.10	34/12.55	68/27.12
diseases	223	238	271	251
	(107-411)	(152-354)	(188-379)	(195-318)

[a] The ratio of observed to expected number of deaths, the standardized mortality ratio and (in parentheses) the 95% confidence interval are shown for each cause of death and duration of employment.

Table 6. Mortality by level of exposure after 20 years since first exposure[a]

Cause of death	Level of exposure	
	Low	High
All causes	122/132.5	204/169.0
	92	121
	(77-110)	(105-138)
All malignant neoplasms	25/26.02	38/34.31
	96	11
	(62-142)	(78-152)
Lung cancer	9/8.42	16/11.41
	107	140
	(49-203)	(80-228)
Respiratory diseases	17/11.43	51/15.69
	149	325
	(87-238)	(242-427)

[a] The ratio of observed to expected numbers of deaths, the standardized mortality ratio and (in parentheses) the 95% confidence interval are shown for each cause of death and level of exposure

Table 7. Mortality by detailed cause: silicotics only[a]

Cause of death[b]	O	E	SMR	95% CI
All causes (000-999)	103	81.42	127	(103-153)
Infectious diseases (000-136):	5	0.54	917	(298-2141)
Tuberculosis of the respiratory system (010-012)	5	0.29	1718	(558-4010)
Other infectious diseases	0	0.25	0	(0-1447)
Malignant neoplasms (140-207):	15	15.11	99	(56-164)
Buccal cavity and pharynx (140-149)	1	0.21	472	(12-2628)
Oesophagus (150)	0	0.28	0	(0-1317)
Stomach (151)	2	2.45	82	(10-295)
Intestine, except rectum (152-153)	1	0.79	127	(3-709)
Rectum (154)	1	0.98	102	(3-570)
Larynx (161)	0	0.20	0	(0-1882)
Trachea, bronchus and lung (162)	9	4.91	183	(84-348)
Leukaemia (204-207)	0	0.38	0	(0-973)
Other neoplasms of the lymphatic and haematopoietic tissue (200-203, 208-209)	0	0.38	0	(0-984)
Other malignant neoplasms	1	4.55	22	(1-123)
Benign and unspecified neoplasms (210-239)	0	0.22	0	(0-1647)
Diseases of the circulatory system (390-458)	34	45.14	75	(52-105)

Table 7 (contd)

Cause of death[b]	O	E	SMR	95% CI
Diseases of the respiratory system (460-519):	41	8.13	504	(362-684)
All other known causes	8	12.26	65	(28-129)
Unknown causes	0	-	-	-

[a] O, observed; E, expected; SMR, standardized mortality ratio; 95% CI, 95% confidence interval
[b] Coded in accordance with the Eighth Revision of the *International Classification of Diseases*

Table 8. Mortality by detailed cause: non-silicotics only[a]

Cause of death[b]	O	E	SMR	95% CI
All causes (000-999)	284	301.16	94	(84-106)
Infectious diseases (000-136):	2	2.17	92	(11-333)
Tuberculosis of the respiratory system (010-012)	0	1.04	0	(0-355)
Other infectious diseases	2	1.13	177	(21-638)
Malignant neoplasms (140-207):	62	61.53	101	(77-129)
Buccal cavity and pharynx (140-149)	2	1.25	160	(19-579)
Oesophagus (150)	0	1.23	0	(0-301)
Stomach (151)	11	8.74	126	(63-225)
Intestine, except rectum (152-153)	2	2.98	67	(8-242)
Rectum (154)	11	3.58	307	(153-550)
Larynx (161)	0	0.85	0	(0-433)
Trachea, bronchus and lung (162)	18	19.80	91	(54-144)
Leukaemia (204-207)	0	1.90	0	(0-194)
Other neoplasms of the lymphatic and haematopoietic tissue (200-203, 208-209)	8	2.16	371	(160-731)
Other malignant neoplasms	10	19.05	53	(25-97)
Benign and unspecified neoplasms (210-239)	0	1.10	0	(0-335)
Diseases of the circulatory system (390-458)	120	139.22	86	(71-103)
Diseases of the respiratory system (460-519):	33	24.62	134	(92-188)
All other known causes	62	72.52	85	(66-110)
Unknown causes	5			

[a] O, observed; E, expected; SMR, standardized mortality ratio; 95% CI, 95% confidence interval
[b] Coded in accordance with the Eighth Revision of the *International Classification of Diseases*

Table 9. Lung cancer mortality by duration of employment in silicotics and non-silicotics after 20 years since first exposure[a]

Subjects	Duration of employment (years)			Total
	1-9	10-19	20+	
Silicotics	0/0.27	3/1.66	6/2.50	9/4.43
	0	181	240	203
	(0-1366)	(37-528)	(88-522)	(93-386)
Non-silicotics	2/3.00	3/4.06	11/8.34	16/15.39
	67	74	132	104
	(8-241)	(15-216)	(66-236)	(59-169)

[a] The ratio of observed to expected numbers of deaths, the standardized mortality ratio and (in parentheses) the 95% confidence interval are shown for silicotics and non-silicotics for each duration of employment

Table 10. Lung cancer mortality by level of exposure in silicotics and in non-silicotics after 20 years since first exposure[a]

Subjects	Level of exposure	
	Low	High
Silicotics	1/0.45	8/3.99
	222	201
	(6-1238)	(87-395)
Non-silicotics	8/7.97	8/7.42
	100	108
	(43-198)	(47-212)

[a] The ratio of observed to expected numbers of deaths, the standardized mortality ratio and (in parentheses) the 95% confidence interval are shown for silicotics and non-silicotics for each level of exposure.

Discussion and conclusions

Although silica dust was long suspected to be a risk factor for lung cancer (Hueper, 1966), no evidence was found until the end of the 1960s of an association between silicosis and lung cancer (Rüttner & Heer, 1969). Autopsy studies have consistently yielded negative results (Parkes, 1982; Reichel, 1976). Neuberger (1981) used data on Austrian silicotuberculotics to show why earlier retrospective studies had failed to provide evidence for an association between dust exposure and lung cancer, arguing that the victims did not live long enough to develop cancer, and lung cancer in particular.

Recent studies provide new evidence of an association between silicosis and lung cancer (Westerholm, 1980; Finkelstein et al., 1982; Gudbergsson et al., 1983; Zambon et al., 1987; IARC, 1987a). An excess of lung cancer has been observed in epidemiological studies of various occupational groups exposed to silica dust (Lynge et al., 1986), foundry workers (Koskela et al., 1976; Fletcher & Ades, 1984; IARC, 1984, 1987b), pottery workers (Thomas & Stewart, 1987), gold miners and workers in talc and firebrick manufacture (Katsnelson et al., 1979). The evaluation of these findings, however, is still controversial (Goldsmith et al., 1982; Heppleston, 1985). In most of these studies, an elevated overall mortality was found, mainly due to non-malignant respiratory diseases and to tuberculosis. Lung cancer mortality has also consistently been found to be in excess.

Our study has been conducted on a population of quarry workers exposed to silica with little or no exposure to other lung carcinogens, and is therefore particularly suitable for testing the hypothesis that a carcinogenic risk from exposure to silica dust exists. Furthermore, the development of silicosis in a subgroup of the study population has made it possible to analyse separately the data for workers receiving compensation for silicosis and those not receiving such compensation.

The main findings of the study are as follows:

(1) The study population has an elevated mortality from diseases of the respiratory system due to deaths from silicosis.

(2) No clear overall excess of lung cancer is discernible. However, when analysed by time since first exposure, by duration of employment and by qualitative exposure level, mortality is consistently 40–50% in excess among workers with the longest follow-up period and duration of employment or highest estimated exposure.

(3) When workers compensated for silicosis are compared with non-silicotics, an almost two-fold increase in lung cancer mortality is seen among silicotics while there is no excess among non-silicotics. This difference is also found after stratifying by time since first exposure, duration of employment, and qualitative exposure level.

In conclusion, our study tends to exclude an elevated risk for lung cancer among workers exposed to silica, but only slightly, if at all, to known carcinogens, while indicating that workers who develop silicosis are at higher risk of developing pulmonary neoplasia. This association, previously reported from studies on silicotics, has not so far been detected in populations exposed to silica.

Since silica has been shown to be carcinogenic in experimental studies (IARC, 1987a), one possible explanation of these findings is that silicosis enhances the carcinogenic properties of silica dust. Its effects appear otherwise to be difficult to detect in humans due to the difficulties in controlling for concomitant carcinogenic exposures and possibly also to the relatively small magnitude of the risk, so that studies with large populations are required in order to achieve sufficient statistical power.

Although an excess risk for lung cancer among silicotics still remains after adjusting for duration of exposure, we cannot rule out the possibility that the excess may be associated with higher levels of exposure, which would then be responsible for both the development of silicosis and of lung cancer. If this is so, a situation of full confounding exists which cannot satisfactorily be resolved.

The slight excess among non-silicotics after 20 years of follow-up and 20 years of employment is very difficult to interpret due to the magnitude of the risk, the low statistical power and the lack of information on smoking.

Acknowledgements

We are indebted for comments on the study design to Dr Tony Fletcher, and for technical assistance to Mrs B. Mundt, Mrs M. Bierlich, Mrs V. Kutzner and Mr B. Müller. The study was supported by the International Agency for Research on Cancer (DEB/85/50).

References

Coleman, M., Douglas, A., Hermon, C. & Peto, J.C. (1986) Cohort study analysis with a Fortran computer program. *Int. J. Epidemiol.*, *15*, 134-137

Finkelstein, M., Kusiak, R., & Suranyi, G. (1982) Mortality among miners receiving workmen's compensation for silicosis in Ontario: 1940-1975. *J. Occup. Med.*, *24*, 663-667

Fletcher, A.C. & Ades, A. (1984) Lung cancer mortality in a cohort of English foundry workers. *Scand. J. Work Environ. Health*, *10*, 7-16

Goldsmith, D.F., Guidotti, T.L. & Johnston, D.R. (1982) Does occupational exposure to silica cause lung cancer? *Am. J. Ind. Med.*, *3*, 423-440

Gudbergsson, H., Kurppa, K., Koskinen, H. & Vasama, M. (1983) An association between silicosis and lung cancer. A register approach. In: *Proceedings of the VIth International Pneumoconiosis Conference, 1983, Bochum, Federal Republic of Germany, 20-23 September*, Vol. 1, Bochum, Bergbau-Genossenschaft, pp. 212-216

Heppleston, A.G. (1985) Silica, pneumoconiosis and carcinoma of the lung. *Am. J. Ind. Med.*, *7*, 285-294

Hueper, W.C. (1966) *Occupational and Environmental Cancers of the Respiratory System (Recent Results Cancer Res.*, Vol. 3), Berlin, Heidelberg, New York, Springer-Verlag

IARC (1984) *IARC Monographs on the Evaluation of the Carcinogenic Risk of Chemicals to Humans. Vol. 34, Polynuclear Aromatic Compounds, Part 3, Industrial Exposures in Aluminium Production, Coal Gasification, Coke Production, and Iron and Steel Founding*, Lyon, International Agency for Research on Cancer, pp. 133-190

IARC (1987a) *IARC Monographs on the Evaluation of the Carcinogenic Risk of Chemicals to Humans. Vol. 42, Silica and Some Silicates*, Lyon, International Agency for Research on Cancer, pp. 39-143

IARC (1987b) *IARC Monographs on the Evaluation of the Carcinogenic Risk of Chemicals to Humans.* Suppl. 7, *Overall Evaluations of Carcinogenicity: An Updating of IARC Monographs Volumes 1 to 42*, Lyon, International Agency for Research on Cancer, pp. 224-225

Katsnelson, B. & Mokronosova, K. (1979) Non-fibrous mineral dusts and malignant tumors. *J. Occup. Med.*, *21*, 15-20

Koskela, R.S., Hernberg, S., Kärävä, R., Järvinen, E. & Nurminen, M. (1976) A mortality study of foundry workers. *Scand. J. Work Environ. Health*, 2, suppl. 1, 73-89

Lynge, E., Kurppa, K., Kristofersen, L., Malker, H. & Sauli, H. (1986) Silica dust and lung cancer: results from the Nordic occupational mortality and cancer incidence registers. *J. Natl Cancer Inst.*, *77*, 883-889

Neuberger, M. (1981) *New Approaches to Risk Estimation of Air Pollutants*, Vienna, Facultas

Parkes, W.R. (1982) *Occupational Lung Disorders*, London, Butterworths

Reichel, G. (1976) The silicoses (anthracosilicoses) [in German] In: Ulmer, W.T. & Reichel, G., eds, *Pneumokoniosen*, Berlin, Springer Verlag, pp. 223-228

Rüttner, J.R. & Heer, H.R. (1969) Silicosis and lung cancer in Switzerland [in German]. *Schweiz. Med. Wschr.*, *99*, 245-249

Thomas, T.L. & Stewart, P.A. (1987) Mortality from lung cancer and respiratory disease among pottery workers exposed to silica and talc. *Am. J. Epidemiol.*, *125*, 35-43

Westerholm, P. (1980) Silicosis. Observation on a case register. *Scand J. Work Environ. Health*, *6*, suppl. 2, 3-86

Zambon, P., Simonato, L., Mastrangelo, G., Winkelmann, R., Saia, B. & Crepet, M. (1987) Mortality of workers compensated for silicosis during the period 1959-1963 in the Veneto Region of Italy. *Scand. J. Work Environ. Health*, *13*, 118-123

Occupational dust exposure and cancer mortality – results of a prospective cohort study

M. Neuberger and M. Kundi

Department of Preventive Medicine, Institute of Environmental Hygiene, University of Vienna, Vienna, Austria

Summary. Over the period 1950–60, occupational and smoking histories were collected in the course of the preventive medical examinations of 247 064 workers in Vienna. Of these, 1630 male workers aged ≥ 40 were selected because of their occupational exposure to silica and 'inert' dusts, and were matched for age, time at which observation was begun, and smoking with 1630 subjects from the same source but without such exposure. Follow-up of 99.8% of the members of these two cohorts resulted in 60 237 person-years of observation, while identification of the underlying cause of death for 98.8% of them (by autopsy in > 50%) up to the end of 1985 showed a significantly higher mortality from lung cancer in dust-exposed subjects (179 cases) as compared with those not so exposed (141 cases) and with the local population (standardized mortality ratio SMR 169). This excess lung cancer mortality was found in all subgroups (SMR in foundries 164, other metal industries 133, ceramics and glass 237, stone and construction 294), consistent with the hypothesis that long-term heavy occupational exposure to silica and 'inert' dusts promotes lung cancer. The only other cancer site for which the number of cases was significantly greater in those exposed to dust was the stomach (SMR 166).

Introduction

Screening examinations were conducted by a mobile team in Vienna during the 1950s on 247 064 workers in 1089 enterprises. On the average, 78% of all active workers were examined (Popper, personal communication). (Some retirees also participated in the screening programme when the health care bus visited their former places of work.) In the course of these preventive examinations, a cohort study was started to test the hypothesis that long-term occupational exposure to high concentrations of silica and 'inert' respirable particulates increases lung cancer mortality. This hypothesis has been the subject of controversy since the negative results of retrospective studies were questioned (Kennaway & Kennaway, 1953; James, 1955; Grosse, 1956). No prospective studies had at that time been carried out. In the present studies, therefore, healthy dust-exposed subjects and matched controls have been followed over a long period.

In earlier reports, details have been given of the source population (Popper & Tuchmann, 1966), the selection of subjects and of multiple-matched controls (Gründorfer & Popper, 1966) and the overall respiratory and cancer mortality related to dust exposure (Neuberger, 1979; Neuberger et al., 1982; Neuberger & Kundi, 1985), special attention being paid to competing causes of death (Neuberger, 1980) and survival (Neuberger et al., 1988) and to showing, with the help of data on Austrian silicotuberculotics (Neuberger, 1979) and silicotics (Neuberger et al., 1988), why earlier studies failed to provide convincing evidence of a relationship between occupational exposure to silicogenic dusts and lung cancer. The present paper summarizes the most important results on cancer mortality and updates them for a further five years, i.e., up to the end of 1985.

Study population

The Vienna occupational health care unit began carrying out preventive medical examinations in 1950; these included both a work history and a smoking history. All subjects with a history of exposure for long periods to high dust (silica) concentrations[1] were chest X-rayed and are the source of our exposed cohort. All men who underwent the first chest X-ray during the period 1950–60 because of their occupational dust (silica) exposure history were eligible for inclusion in the study if they were born before 1911 (and were therefore aged ≥ 40 years when the observations were begun) and were resident in Vienna ($n=1630$). The age distribution at the date of inclusion is shown in Table 1. An equal number of male Viennese subjects without occupational dust exposure were selected from the same examination records by matching year of birth (age), year of first examination (beginning of observation) and smoking history (smoker or non-smoker at the time of the first examination); these subjects served as a control cohort.

The two cohorts constitute a sample of elderly (most were born between 1896 and 1904) Viennese workers surveyed in the 1950s. Those exposed to dust were employed mainly in the metal industry (e.g., grinders), almost half in iron foundries (casters, melters, core-makers, fettlers, etc.), while smaller numbers worked in the ceramics, glass and stone industry and in certain other branches of industry (e.g., production of sand, cleaning agents, etc.). The unexposed workers were employed in a large variety of industries (publishing and printing, chemicals, construction, electrical engineering, textiles, leather, food, etc.). Since exposure to dust dates back to the 1930s and in some cases even to the beginning of the century, no personal exposure data were available. However, in each case, examination not only of the enterprise, department and job title, but also of the personal history, showed that exposure to dust had taken place, while those known or suspected to have been exposed to asbestos (e.g., insulators) were excluded.

---

[1] A minimum of five years' exposure (in most cases subjects were exposed for the greater part of their working lives) to more than 6 mg/m^3 respirable dust.

Table 1. Age distribution of subjects at time of inclusion in the study

Age (years)	Exposed		Controls	
	n	%	n	%
40-44	139	8.5	139	8.5
45-49	341	21.0	341	21.0
50-54	442	27.2	442	27.2
55-59	400	24.6	400	24.6
60-64	228	14.0	228	14.0
65-69	59	3.6	59	3.6
70-74	13	0.8	13	0.8
75-80	4	0.2	4	0.2
Total	1626	100.0	1626	100.0

Methods

Systematic inquiries at the registrar's and public health offices have made it possible to trace both exposed and unexposed subjects up to the end of 1985. The most recent data obtained from the registrar's office, which cover the period 1981–85, has been checked against additional information[1]. Primary and secondary causes of death were taken from official death certificates and, if necessary, discussed with the reporting pathologist, hospital or physician. (The autopsy rate in Vienna exceeds 50% and rises to 57% for malignant neoplasms). All diagnoses were traced and coded without knowledge of the subject's exposure to dust. In order to compare the causes of death among the cohort with those in the general population, the strategy and coding scheme used to determine the primary (underlying) cause of death was the same as that employed by the statistical office. Best estimates of the causes of death were included as well as secondary causes of death for use in comparisons between cohorts. In this paper, however, only cause-specific and overall SMRs based on underlying cause will be given. The expected number of deaths computed is based on the general Viennese population; this will give conservative estimates, since mortality in Vienna, and especially cancer mortality, is higher than in the rest of Austria and some of the subjects moved to rural areas after retirement. Statistical significance was assessed by means of chi square tests.

---

[1]The time taken for deaths to be recorded in the files at the Vienna registrar's office varies. In particular, it is possible that subects recorded in Vienna as alive have left the city and died elsewhere, so that registers for other areas must also be checked. There is no uncertainty, however, about those reported to be dead, while a multistage checking procedure has been used to confirm that those reported to be alive have been correctly registered.

Results

A total of 1626 exposed and 1626 unexposed subjects (99.8%) were traced up to the end of 1985 or until death, giving a total of 28 972 person-years of observation in the exposed group and 31 265 person-years in the control group. Of those exposed, 1418 (87%) were recorded as having died; the corresponding figures for the unexposed were 1364 and 84%. Five subjects (two exposed and three unexposed) who had gone abroad were excluded at the date of emigration. The cause of death could not be identified for 16 subjects (seven exposed and nine unexposed) but was ascertained for 99%.

The SMR for all causes of death (Table 2) was found to be 114 for those exposed to dust, which is statistically significantly greater than that for the Viennese reference population, and 94 for the controls, which is less than the reference value. Since the youngest of those still alive at the end of 1985 were already 75 years old, these figures are not likely to change with further follow-up

Table 2. All deaths, cancer deaths and standardized mortality ratios, 1950–85

Cancer site and ICD code[a]	Exposed (n=1626)		Controls (n=1626)	
	No.	SMR	No.	SMR
Lung (ICD 162)	179	169[b]	141	118[c]
Stomach (ICD 151)	77	166[b]	47	90
Intestine (ICD 152-154)	34	81	50	102
Other digestive organs (ICD 140-150, 155-159)	48	117	42	90
Other sites	71	88	88	94
All deaths	1418	114[b]	1364	94[c]

[a] From the Eighth Revision of the *International Classification of Diseases*
[b] $p < 0.01$
[c] $p < 0.05$

investigations. The excess mortality in workers exposed to dust is due mainly to the following causes of death: lung cancer (SMR 169), stomach cancer (SMR 166), emphysema, chronic bronchitis and bronchial asthma (SMR 207), silicosis and silicotuberculosis. A thorough investigation of the reasons for the lower mortality of the controls showed that it could be attributed to the 'healthy worker' effect, since it is significant only below age 60 and up to ten years after recruitment into the study.

As shown in Table 2, we found a substantially higher mortality from lung and stomach cancer in the exposed group, but no indication of excess mortality from any other cancers. In the controls, the mortality pattern for cancers was the same as in the general Viennese population, except for a slight increase in lung cancer (SMR 118).

Table 3 shows cancer deaths and SMRs stratified according to the industry recorded in the entry files of those exposed to dust (except for 33 workers with ill-defined job titles or several jobs in the entry file). This stratification of the exposed group shows that there is an excess mortality from lung cancer in all subgroups. The SMR for stomach cancer was greater than 100 for all strata, but statistically significant only for those employed in the metal industry.

Table 3. Cancer deaths and standardized mortality ratios by industry

Cancer site	Foundries ($n=775$)		Other metal ($n=475$)		Ceramics and glass ($n=191$)		Stone and construction ($n=87$)		Other ($n=65$)	
	No.	SMR	No.	SMR	No.	SMR	No.	SMR	No.	SMR
Lung	85	164[a]	43	133[b]	28	237[a]	15	294[a]	6	149
Stomach	40	177[a]	22	158[b]	6	116	4	174	4	222
Intestine	15	74	7	53[b]	7	150	2	101	3	189
Other digestive organs	19	95	19	151[b]	5	108	2	102	2	130
Other sites	30	77	26	104	6	67	5	132	2	66

[a] $p < 0.01$
[b] $p < 0.05$

Discussion

A high lung cancer mortality among workers exposed to silica and 'inert'[1] dusts has been reported (e.g., Koskela *et al.*, 1976; Gibson *et al.*, 1977; Selikoff, 1978; Palmer & Scott, 1981; Katsnelson & Mokronosova, 1979; Goldsmith *et al.*, 1982; Thomas, 1982; Fletcher & Ades, 1984). However, the results of the present study, as reported from 1966 onwards, have consistently shown a higher lung (and stomach) cancer mortality in iron foundry workers as well as in other workers exposed to dust. The higher lung cancer mortality in foundry workers has been attributed by some authors to the fact that they are exposed to polycyclic aromatic hydrocarbons (PAH), but it is of the same magnitude as that of other workers exposed to dust but not occupationally exposed to PAH.

The relationship between the increased lung cancer risk in iron foundries and PAH was found not to be significant by Tola *et al.* (1979), and in a steel foundry only the high risk for crane-drivers was found to be correlated with high PAH levels. Benzo[*a*]pyrene levels were found to be higher in areas used for moulding and around furnaces where lung cancer mortality was lower (Gibson *et al.*, 1977).

---

[1]These are dusts (deposited in the airways and lungs) which would be expected not to produce any specific (toxic, fibrogenic, carcinogenic, etc.) reaction but only effects on respiratory clearance and carrier effects.

In contrast, increased lung cancer rates were found in the finishing area, furthest away from the source of PAH and where levels of benzo[a]pyrene were lowest (but there were high dust exposures from grinding and sand-blasting). In another study, fettlers and maintenance workers exposed to low levels of PAH and high levels of dust were identified as persons with an increased lung cancer risk (Fletcher & Ades, 1984).

We would suggest that the above-mentioned results should be seen in the context of a multistage model of carcinogenesis. Genotoxic PAH from foundry air as well as from tobacco smoke (adsorbed on to particles) might act as initiators and the particles as promoters. With synergism of this kind, lung cancer might well increase with increasing dust load and not necessarily with PAH levels, even though both might be involved. The information available on other possible exposures (Neuberger et al., 1988) provides no evidence to show that PAH might have played an important role as confounders in our study. Controls were drawn from the same source population, the only requirement being the absence of any history of substantial exposure to dust (or silica); those exposed to gases and vapours, solvents, oils, dyes, heavy metals, etc., were therefore not excluded. Of course, there are differences in exposure between e.g., a typesetter, lead caster or galvanizer in the control cohort and a foundry worker, iron caster or grinder in the exposed cohort. Similarly, in the construction industry, there are differences between the exposures of a road worker (working with tar) or roofer (possibly exposed to asbestos) in the control cohort and a stone cutter or tunnel worker (possibly exposed to radon) in the exposed cohort. Even workers at the same plant (e.g., a stone-cutter exposed to dust and a driver as a control) will have very different exposures. There are more metal workers and fewer construction workers in the exposed cohort, and some branches of industry (e.g., printing and publishing, the film industry, the food industry, etc.) are represented only in the control cohort, since workplaces in which employees were exposed to dust do not exist in them.

In both the dust-exposed and the control cohort, however, subjects have been occupationally exposed to gases, vapours, solvents, oil, tars, dyes, heavy metals, etc., as well as to carcinogens such as PAH. The lung cancer excess in the non-dust-exposed cohort (SMR 118) might be the consequence of such exposures, and this excess must be deducted from the excess observed in the dust-exposed workers (SMR 169) before the dust-related lung cancer can be estimated from a comparison with the general population. More important, however, is the comparison between dust-exposed and non-dust-exposed workers matched for smoking, which gave a relative lung cancer risk of 1.4. This lung cancer risk can be related only to the dust exposure. Total respirable particulates and silica both greatly exceeded the maximum laid down in the standards currently in force (Neuberger et al., 1982).

Confounding exposures after entry into the study have not been controlled, but seem to be of minor importance because of the marked tendency of the workers concerned to remain in the same industry and even in the same plant (especially after age 40). In addition, the pensions insurance board provided a complete list of all the jobs held by and places of work of dust-exposed lung cancer cases who died up to 1980; this confirmed the stability of the workforce and the reliability of the work histories and failed to show any second jobs with exposures to human carcinogens such as asbestos, chromates, nickel, etc., up to retirement.

The possible occurrence of confounding exposures cannot, however, be disregarded, because in obtaining the work histories by interview at the beginning of our cohort study, we relied mainly on the subject's memory (Popper & Tuchmann, 1966) and exposures thereafter could be found only from insurance records.

The increased lung cancer risk of those exposed to dust did not become apparent before the age of 60 (Neuberger et al., 1986). A detection bias from the X-ray screening of such subjects can therefore be disregarded, since it is discontinued when exposure ceases or on retirement, perhaps fortunately for our study but unfortunately for the workers.

Various selection biases (Neuberger & Raber, 1983) can lead to a distortion of the results, mainly in studies on silicotics (e.g., selection because of premature death from other causes), but must also be taken into account in prospective cohort studies. Choice of and employment in a dusty job (influenced, e.g., by health status and other requirements, such as the prohibition of smoking in coal mines), selection out of the job (e.g., by respiratory disease), and competing causes of death (premature deaths and deaths from multiple causes) can preclude the detection of an increased lung cancer risk in dust-exposed cohorts. Because of the higher age of entry into our cross-sectional cohort study (healthy worker selection) and because of the death rates from competing causes (selection by death due to non-malignant respiratory diseases), it is possible that, in our cohort, the true lung cancer risk related to dust exposure has been underestimated.

Initially, our study was based on the hypothesis that chronic irritation of the bronchial mucosa by respirable non-fibrous particulates increases lung cancer incidence. Since then, other and more detailed explanations of the pathogenesis of lung cancer have been developed, taking into account the combined effects of silica (Goldsmith et al., 1986), 'inert' dust exposure (Higgins et al., 1979) and smoking on mucociliary and alveolar clearance. Our findings are also compatible with the hypothesis that the long-term bombardment of the airways by dust particles, even by inert ones, by maintaining high levels of polymorphonuclear recruitment and alveolar macrophage activation, causes lung cancer by mechanism which - acting ultimately at the cellular level - are the same for all inhaled particles regarded as inert, or at least non-carcinogenic (Becklake, 1985).

Elevated rates of stomach cancer in certain dusty trades have been ascribed to carcinogens adsorbed on to particles, cleared from the lung, swallowed and then acting on the gastric mucosa (Meyer et al., 1980), but no precise pathogenic mechanisms at the cellular level have been defined and studies on a number of different dust-exposed cohorts have not given consistent results. As the focus of our study was the investigation of the lung cancer risk, confounding factors for stomach cancer such as dietary habits, have not been controlled. For this reason, great care must be taken in assessing the significance of the increased mortality from stomach cancer in our dust-exposed cohort, although our results are consistent with the hypothesis mentioned above.

Acknowledgements

This research was funded by a grant from the Medical Science Fund of the Mayor of the City of Vienna. The authors are grateful to Dr W. Gründorfer and the late Professor Dr L. Popper for providing the data base, to the registration, registrar's and public health offices for help with the follow-up, to the statistical office for providing demographic data, to the workers' pensions insurance and the workers' compensation board for additional information on work histories after registration and to the Biostatistics Unit of IARC for helpful suggestions.

References

Becklake, M.R. (1985) Chronic airflow limitation: its relationship to work in dusty occupations. *Chest*, 88, 608-617

Fletcher, A.C. & Ades, A. (1984) Lung cancer mortality in a cohort of English foundry workers. *Scand. J. Work Environ. Health*, 10, 7-10

Gibson, E.S. Martin, R.H. & Lockington, J.N. (1977) Lung cancer mortality in a steel foundry. *J. Occup. Med.*, 19, 807-812

Goldsmith, D.F., Guidotti, T.L. & Johnston, D.R. (1982) Does occupational exposure to silica cause lung cancer? *Am. J. Ind. Med.*, 3, 423-440

Goldsmith, D.F., Winn, D.M. & Shy, C.M., eds (1986) *Silica, Silicosis and Cancer: Controversy in Occupational Medicine* (Cancer Research Monographs, Vol. 2), New York, NY, Praeger

Grosse, H. (1956) Silicosis and lung cancer [in German]. *Arch. Gewerbepathol. Gewerbehyg.*, 14, 357-372

Gründorfer, W. & Popper, L. (1966) Dust exposure and incidence of bronchial carcinoma [in German]. In: *Proceedings, XV International Congress on Occupational Health, Vienna, 19-24 September 1966*, Vol. 3, Vienna, Wiener Medizinische Akademie, pp. 173-176

Higgins, I.T., Albert, R.E., Charlson, R.J., Darley, E.F., Ferris, B.G., Frank, R., Whitby, K.T. & Redmond, J. (1979) *Airborne Particles*, Baltimore, MD, University Park Press

James, W.R.L. (1955) Primary lung cancer in South Wales coalworkers with pneumoconiosis. *Br. J. Ind. Med.*, 12, 87-91

Katsnelson, B.A. & Mokronosova, K.A. (1979) Nonfibrous mineral dusts and malignant tumors: an epidemiological study of mortality. *J. Occup. Med.*, 21, 15-20

Kennaway, E.L. & Kennaway, N.M. (1953) The incidence of cancer of the lung in coal miners in England and Wales. *Br. J. Cancer*, 7, 10-18

Koskela, R.-S., Hernberg, S., Kärävä, R., Järvinen, E. & Nurminen, M. (1976) A mortality study of foundry workers. *Scand. J. Work Environ. Health*, 2, suppl. 1, 73-89

Meyer, M.B., Luk, G.D., Sotelo, J.M., Cohen, B.H. & Menkes, H.A. (1980) Hypothesis: The role of the lung in stomach carcinogenesis. *Am. Rev. Respir. Dis.*, 121, 887-892

Neuberger, M. (1979) *New Approaches to Risk Estimation of Air Pollutants*, Vienna, Facultas

Neuberger, M. (1980) Causes of death of dust workers. Results of a prospective study and their implications for prevention [in German]. In: Brenner, W., Rutenfranz, J., Baumgartner, E. & Haider, M., eds, *Proceedings of the German Association for Occupational Medicine*, Stuttgart, Gentner, pp. 317-323

Neuberger, M. & Raber, A. (1983) Different selection rates - fallacies in epidemiological studies [in German]. *Zbl.-Bakt.-Hyg. B*, 177, 539-561

Neuberger, M. & Kundi, M. (1985) Health risks caused by occupational dust exposure [in German]. *Staub Reinhalt. Luft*, 45, 131-135

Neuberger, M., Kundi, M., Haider, M. & Gründorfer, W. (1982) Cancer mortality of dust workers and controls - results of a prospective study. In: *Prevention of Occupational Cancer - International Symposium*, Geneva, International Labour Office (Occupational Safety and Health Series No. 46), pp. 235-241

Neuberger, M., Kundi, M., Westphal, G. & Gründorfer, W. (1986) The Viennese dusty worker study. In: Goldsmith, D.F., Winn, D.M. & Shy, C.M., eds, *Silica, Silicosis and Cancer: Controversy in Occupational Medicine* (Cancer Research Monographs, Vol. 2), New York, NY, Praeger, pp. 415-422

Neuberger, M., Westphal, G. & Bauer, P. (1988) Long term effect of occupational dust exposure. *Jpn. J. Ind. Health*, *30*, 362-370

Palmer, W.G. & Scott, W.D. (1981) Lung cancer in ferrous foundry workers - a review. *Am. Ind. Hyg. Assoc. J.*, *42*, 329-340

Popper, L. & Tuchmann, E. (1966) Occupational health care of the Viennese health insurance for labourers and employees [in German]. In: *Proceedings, XV International Congress on Occupational Health, Vienna, 19-24 September 1966*, suppl. II, Vienna, Wiener Medizinische Akademie, pp. 10-75

Selikoff, I.J. (1978) Carcinogenic potential of silica compounds. In: Bendz, G. & Lundquist, I., eds, *Biochemistry of Silicon and Related Problems*, New York, Plenum Press, pp. 311-336

Thomas, T.L. (1982) A preliminary investigation of mortality among workers in the pottery industry. *Int. J. Epidemiol.*, *11*, 175-180

Tola, S., Koskela, R.-S., Hernberg, S. & Järvinen, E. (1979) Lung cancer mortality among iron foundry workers. *J. Occup. Med.*, *21*, 753-760

Lung cancer mortality among pottery workers in the United States

T. L. Thomas

Department of Veterans Affairs, Veterans Health Services and Research Administration, Office of Environmental Epidemiology, Washington, DC, USA

Summary. A proportionate mortality study suggested that members of the International Brotherhood of Potters and Allied Workers in the United States had an elevated frequency of deaths from non-malignant respiratory disease (PMR=1.54) and lung cancer (PMR=1.21). The lung cancer excess occurred exclusively among pottery workers employed in the manufacture of plumbing fixtures (PMR=1.80). A subsequent cohort study examined mortality among 2055 white men employed in three ceramic plumbing fixture factories. There was a significant excess of non-malignant respiratory disease (SMR=1.73). Lung cancer mortality was also higher than expected (SMR=1.43) and was highest among workers whose jobs involved simultaneous exposure to silica and non-fibrous talc (SMR=2.54). Lung cancer mortality risk increased with increasing number of years of exposure to non-fibrous talc and showed no pattern by number of years of exposure to silica. Among men exposed to talc, lung cancer risk increased with years since first non-fibrous talc exposure and decreased with age at first exposure. The data suggested an association between exposure to non-fibrous talc and excess lung cancer risk; however, the role of silica as a co-factor or promoting agent could not be ruled out.

Introduction

Investigations of the relationship between occupational exposure to silica and lung cancer were begun by the National Cancer Institute in 1976 after an inquiry by a physician who had treated several lung cancer patients employed in a pottery plant that produced plumbing fixtures. Because no studies of the long-term health effects of silica on pottery workers had been conducted in the United States since the 1930s, an investigation of mortality among workers in this industry was initiated using the records of the International Brotherhood of Potters and Allied Workers (Thomas, 1982). A proportionate mortality ratio (PMR) study was conducted as a pilot phase.

Records of death benefits paid to relatives of active and retired members of the International Brotherhood of Potters and Allied Workers between 1955 and 1977 were used to identify members of the study group. Death certificates were obtained for 2924 white male decedents from state vital records offices and the underlying cause of death was classified by a qualified nosologist. Cause-specific PMRs were calculated by comparing observed deaths in the study group with those expected based on the mortality experience of white men in the general population of the United States, with appropriate adjustments for age at death and calendar period (Monson, 1974).

Excesses of mortality from lung cancer and non-malignant respiratory disease were seen among all white male union members (Table 1). When the data were examined by the type of product manufactured, the lung cancer mortality excess was confined to men who had worked in the manufacture of ceramic plumbing fixtures (sanitary ware), while the respiratory disease excess occurred among men who had worked in all types of pottery manufacture. Crystalline silica dust is the substance to which sanitary-ware workers are mainly exposed, but they may also be exposed to talc dust, soda ash, borate minerals, fluorspar, phosphate minerals, pigments and numerous substances used in glazes, including antimony, chromium, copper, iron, titanium and many others. A cohort mortality study was undertaken to further investigate the risk of non-malignant respiratory disease and lung cancer among workers in the plumbing-fixture industry (Thomas & Stewart, 1987).

Table 1. Proportionate mortality from lung cancer among 2924 white male pottery workers in the United States by type of product

Product (total deaths)	Lung cancer			Non-malignant respiratory disease		
	Observed deaths	Expected	PMR	Observed deaths	Expected	PMR
All products ($n=2924$)	178	146.6	1.21^a	268	173.7	1.54^a
Sanitary ware ($n=573$)	62	34.4	1.80^a	43	30.4	1.41
Other products ($n=2351$)	116	112.2	1.03	225	143.3	1.57^a

$^a p < 0.01$

Materials and methods

Three factories of a single United States company producing ceramic plumbing fixtures were selected for study. Company personnel rosters and files were used to identify 2055 white male employees who had completed at least one year of employment with the company between 1 January 1939 and 1 January 1966. The vital status of each study subject on 1 January 1981 was determined from company, Social Security Administration, credit bureau, and department of motor vehicles records. Of the men in the study group, 68% were known to be alive, 28% were deceased, and for only 4% was the vital status unknown. Underlying causes of death were obtained from the death certificates of subjects known to be deceased. Standardized mortality ratios (SMRs) were calculated as the ratio of deaths observed in the study group to those expected based on the mortality of

United States white men, with appropriate adjustment for age and calendar period; statistical significance was determined at the 5% level using a chi square test with one degree of freedom (Monson, 1974).

The manufacture of sanitary ware in the United States involves the mixing of the raw materials (slip making), the casting of the liquid clay into plaster-of-Paris moulds, finishing, glazing, firing, inspection and packaging (Thomas & Stewart, 1987). As already mentioned, exposure to crystalline-free silica, which is a major constituent of the raw materials, is the main exposure in numerous factory departments. In the three factories studied, non-fibrous Montana steatite talc has been used to dust moulds for casting since about 1955 and is believed to contain no asbestiform fibres (Pfizer, personal communication, 1985; Grexa & Parmentier, 1979). Before 1955, flint and ground clay were used to dust moulds. Glazes consist of raw materials similar to those making up the liquid clay. In the past, tremolitic (fibrous) talc was used in some glazes, but its use was discontinued in these factories in 1976.

Based on a detailed knowledge of industrial processes and job duties, each job title-department combination listed in the work history of individual study subjects was classified according to its potential exposure to silica dust (none, low, or high) and to talc (none, non-fibrous, or fibrous) (Thomas & Stewart, 1987). A hierarchy for exposure to silica and talc was created from these classifications. Jobs not involving exposure to silica or talc were placed in the lowest category in the hierarchy, immediately below jobs with low silica exposure. Exposure to talc occurred exclusively in work areas where there was also high silica exposure; the high silica exposure category was therefore subdivided in accordance with the talc exposure, no talc being rated the lowest and fibrous talc the highest. Thus, the lowest category contained person-years during which a subject had never been exposed to silica or talc, while the highest contained all person-years after which a subject had been exposed to fibrous talc.

Results

The number of deaths observed for all cancers was nearly the same as expected, but there was a significant excess of lung cancer deaths (Table 2). The SMR for non-malignant respiratory disease was also significantly elevated. This finding was attributed to a nearly three-fold excess of deaths from non-malignant respiratory disease not classified as pneumonia or emphysema, which included 23 deaths from what were classified as pneumoconioses due to silica and silicates.

Excess risks of lung cancer and non-malignant respiratory disease were observed exclusively among persons exposed to high levels of silica dust (Table 3). SMRs for non-malignant respiratory disease were consistently elevated among workers both with and without exposure to non-fibrous talc but not among those exposed to fibrous talc. Lung cancer mortality was slightly elevated both among persons with exposure to fibrous talc and those with no talc exposure, but there was a significant 2½-fold excess among workers with exposure to non-fibrous talc. Mortality from non-malignant respiratory disease increased with duration of silica exposure, but lung cancer mortality did not (Thomas & Stewart, 1987). Mortality from lung cancer increased, but mortality from non-malignant respiratory disease decreased with increasing duration of non-fibrous talc exposure (Thomas & Stewart, 1987).

Table 2. Mortality 1940-80, among 2055 white male sanitary-ware workers

Underlying cause of death	Observed deaths	Expected deaths	SMR[a]
All causes of death	578	645.1	0.90[b]
Malignant neoplasms:	124	122.0	1.02
Lung cancer	52	36.3	1.43
Non-malignant respiratory disease	64	37.0	1.73[b]
Pneumonia	16	14.3	1.12
Emphysema	7	8.5	0.82
Other respiratory disease	41	14.1	2.90[b]

[a] SMR, standardized mortality ratio
[b] $p < 0.05$

Table 3. Lung cancer and non-malignant respiratory disease mortality, 1940-80, among white male sanitary-ware workers by exposure category

Exposure category	Lung cancer		Non-malignant respiratory disease	
	Observed deaths	SMR[a]	Observed deaths	SMR[a]
No silica, no talc	1	0.61	1	0.55
Low silica, no talc	7	0.68	9	0.81
High silica:	44	1.81[b]	54	2.26[b]
No talc	18	1.37	36	2.64[b]
Non-fibrous talc	21	2.54[b]	16	2.20[b]
Fibrous talc	5	1.74	2	0.67

[a] SMR, standardized mortality ratio
[b] $p < 0.05$

Patterns of lung cancer mortality were examined in detail and separately in relation to non-fibrous talc exposure and to silica exposure in order to explore further the relationship between the two substances and lung cancer mortality risk. Table 4 shows results for workers with simultaneous exposure to high silica levels and non-fibrous talc. Mortality from lung cancer rose with increasing duration of non-fibrous talc exposure to a statistically significant SMR of 3.64 after 15 years. Lung cancer mortality risk was significantly elevated after five years since first non-fibrous talc exposure, but did not increase with time thereafter. Among men first exposed to non-fibrous talc before age 40, lung cancer mortality risk was significantly elevated and decreased slightly with age at first exposure. Among terminated and retired workers, the risk decreased slightly ten years after exposure to non-fibrous talc had ceased, but there were very few terminated employees who were followed for more than 15 years after exposure had ended.

Table 4. Lung cancer mortality, 1940–80, among white male sanitary-ware workers exposed simultaneously to non-fibrous talc and high silica levels

	Observed deaths	SMR[a]
Total	21	2.54[b]
Years exposed to non-fibrous talc:		
< 5	2	0.95
5–14	11	2.76[b]
15+	8	3.64[b]
Years since first non-fibrous talc exposure:		
< 5	0	0.00
5–14	8	2.81[b]
15+	13	2.75[b]
Age at first non-fibrous talc exposure:		
< 40	9	3.51[b]
40–49	6	2.17
50+	6	2.03
Years since last non-fibrous talc exposure (terminated workers only):		
5	7	3.83[b]
5–9	5	3.76[b]
10+	3	2.05

[a] SMR, standardized mortality ratio
[b] $p < 0.05$

Table 5 shows detailed results for workers who had at any time been exposed to silica. Among all silica-exposed workers, the lung cancer risk was lowest among those who had the longest duration of silica exposure. There was no pattern of increasing lung cancer mortality with latency period, but the SMR was statistically significant among men who had survived for 30 years since their first exposure. The lung cancer mortality risk decreased with age first exposed and showed no pattern in relation to the number of years since last exposure among terminated workers. When the analyses were restricted to men never exposed to talc, the lung cancer mortality risk increased slightly with the number of years since first exposure, but decreased with increasing duration of silica exposure. The risk did not change markedly with age at first silica exposure. There was a three-fold excess of lung cancer mortality among men who had terminated their exposure more than 30 years before their death. These patterns did not change when workers exposed only to low silica levels were excluded from the analyses.

Table 5. Lung cancer mortality, 1940–80, among white male sanitary-ware workers exposed to silica at any time

	All workers		Workers never exposed to talc	
	Observed deaths	SMR[a]	Observed deaths	SMR[a]
Total	51	1.47[b]	25	1.09
Years exposed to silica:				
< 15	19	1.62	14	1.50
15–29	19	1.68[b]	7	0.99
30+	13	1.12	4	0.62
Years since first silica exposure:				
< 15	4	1.46	1	0.60
15–29	15	1.32	4	0.55
30+	32	1.56[b]	20	1.43
Age at first silica exposure:				
< 40	45	1.51[b]	22	1.11
40+	6	1.25	3	0.97
Years since last silica exposure (terminated workers only):				
< 15	29	2.10[b]	12	1.40
15–29	6	0.86	5	0.85
30+	7	3.08	7	3.19[b]

[a] SMR, standardized mortality ratio
[b] $p < 0.05$

Discussion

The excess mortality from lung cancer in this study occurred primarily among pottery workers who were simultaneously exposed occupationally to non-fibrous talc and silica; the roles of the two agents cannot therefore be entirely separated. Occupational exposure to silica is known to cause silicosis, a chronic fibrotic lung condition (Jones, 1983). If lung cancer and chronic lung disease had an etiological factor in common, the two diseases might be expected to exhibit similar patterns. However, in the present study, lung cancer mortality and respiratory disease mortality showed markedly different patterns by calendar period of first employment, duration of exposure to silica, and time since first exposure to silica (Thomas & Stewart, 1987).

Among men exposed only to silica, the lung cancer risk decreased with increasing duration of silica exposure and showed no relationship with age at first exposed or the number of years since silica exposure ended. Except for a slightly elevated lung cancer SMR that was not statistically significant among men who survived for 30 years after their first exposure, the evidence presented provides little support for a causal relationship between silica exposure alone and lung cancer.

Among men with simultaneous exposure to non-fibrous talc and silica, patterns of increasing risk with duration of non-fibrous talc exposure, the significantly elevated lung cancer mortality risk after five years since first non-fibrous talc exposure, and decreasing risk by age at first non-fibrous talc exposure are similar to those suggested for an early-stage carcinogen (Day & Brown, 1980; Chu, 1987). The slightly decreasing risk with the number of years since last non-fibrous talc exposure could suggest, on the other hand, that non-fibrous talc might act as a late-stage carcinogen, but the numbers of workers who had ceased to be exposed more than 15 years previously was too small to make it possible to determine whether the risk would fall to the background level. Few studies of workers occupationally exposed to talc have been conducted to date; however, three studies of steatite talc miners showed an elevated risk of lung cancer (Katsnelson & Mokronosova, 1979; Kleinfeld et al., 1974; Selevan et al., 1979) and analyses of the talc indicated that it was not contaminated with asbestos and contained only very small amounts of quartz (Boundy et al., 1979; Katsnelson & Mokronosova, 1979). Since the non-fibrous talc exposure in this study did not begin until 1955, further studies should focus on workers with longer exposure and latency periods. The evidence suggests that exposure to non-fibrous talc is related to the excess lung cancer risk; however, the possibility that silica acts as a co-factor or promoting agent cannot be ruled out.

References

Boundy, M.G., Gold, K., Martin, K.P., Burgess, W.A. & Dement, J.M. (1979) Occupational exposure to non-asbestiform talc in Vermont. In: Lemen, R. & Dement, J.M., eds, *Dust and Disease*, Park Forest South, IL, Pathotox Publishers, pp. 365-378

Chu, K.C. (1987) A nonmathematical view of mathematical models for cancer. *J. Chron. Dis.*, *40*, suppl. 2, 163S-170S

Day, N.E. & Brown, C.C. (1980) Multistage models and primary prevention of cancer. *J. Natl Cancer Inst.*, *64*, 977-989

Grexa, R.W. & Parmentier, C.J. (1979) Cosmetic talc properties and specifications. *Cosmet. Toilet.*, *94*, 29-33

Jones, R.N. (1983) Silicosis. In: Rom, W.N., ed. *Environmental and Occupational Medicine*, Boston, MA, Little, Brown, pp. 197-205

Katsnelson, B.A. & Mokronosova, K.A. (1979) Nonfibrous mineral dusts and malignant tumors: an epidemiological study of mortality. *J. Occup. Med.*, *21*, 15-20

Kleinfeld, M., Messite, J. & Zaki, M.H. (1974) Mortality experiences among talc workers: a follow-up study. *J. Occup. Med.*, *16*, 345-349

Monson, R.R. (1974) Analysis of relative survival and proportional mortality. *Comput. Biomed. Res.*, *7*, 325-332

Selevan, S.G., Dement, J.M., Wagoner, J.K. & Froines, J.R. (1979) Mortality patterns among miners and millers of non-asbestiform talc. In: Lemen, R. & Dement, J.M., eds, *Dust and Disease*, Park Forest South, IL, Pathotox Publishers, pp. 379-388

Thomas, T.L. (1982) A preliminary investigation of mortality among workers in the pottery industry. *Int. J. Epidemiol.*, *11*, 175-180

Thomas, T.L. & Stewart, P.A. (1987) Mortality from lung cancer and respiratory disease among pottery workers exposed to silica and talc. *Am. J. Epidemiol.*, *125*, 35-43

A mortality follow-up study of pottery workers: preliminary findings on lung cancer

P.D. Winter[1], M.J. Gardner[1], A.C. Fletcher[1] and R.D. Jones[2]

[1]*Medical Research Council Environmental Epidemiology Unit, (University of Southampton), Southampton General Hospital, Southampton, England*
[2]*Epidemiology and Medical Statistics Unit, Health and Safety Executive, Bootle, Merseyside, England*

Summary. The possible association between exposure to low levels of silica and lung cancer was investigated by following up pottery workers included in a survey conducted in 1970–71 of respiratory disease among such workers. The initial results show that, among men under the age of 60 at the time of the original survey, mortality has been similar to that expected, but that there was an excess of lung cancer of over 30% even after allowance had been made for cigarette smoking and place of residence. There were no particular excesses of lung cancer by product group or job group. However, there was some indication that lung cancer risk increased with estimated cumulative exposure to respirable quartz. These findings do suggest an association between lung cancer and the low levels of silica found in potteries, and the follow-up will therefore be continued and a more detailed analysis of the data carried out.

Introduction

Several studies have reported that persons with silicosis have a higher incidence of lung cancer as compared with the general population and this has led to the conclusion that silica dust is probably a human carcinogen (IARC, 1987). There is also concern that there may be a dose–response relationship without a safe threshold, and hence, potentially, a risk of lung cancer from lower levels of silica exposure than those associated with recognized silicosis.

During 1970–71 the Health and Safety Executive (HSE) carried out a survey of respiratory disease among a stratified sample of the then current workforce of the pottery industry in Britain (Fox *et al.*, 1975). A range of different potteries was studied, and the data collected on each individual included respiratory function, previous occupational history and cigarette smoking habits, among other factors. Industrial hygiene measurements were also undertaken at the time, and the majority of samples were below the then current threshold limit value (0.1 mg/m^3).

It was recognized that the follow-up of these employees over the ensuing years would enable information to be obtained to help answer questions about the health effects of relatively low exposures to silica. This paper describes the procedure used in this follow-up and gives some preliminary findings.

Methods

At the outset, an attempt was made through the HSE to trace the original questionnaires and paper records of the workers included in their survey. This, unfortunately, was unsuccessful. However, a computer tape with the coded survey data was located, containing, in particular, the National Insurance number and date of birth of each worker, and the follow-up study was mounted from these identification particulars.

This practice is unusual for cohort studies in the United Kingdom, which would normally be mounted using full names, dates of birth and National Health Service numbers, if available. These particulars make it possible to link up directly with the National Health Service Central Register (NHSCR), which contains information on vital status and cause of death. However, in this instance, it was first necessary to use the National Insurance register held by the Department of Health and Social Security (DHSS) to obtain further identification of the individual pottery workers.

The basic intention was first to obtain the full names of all the surveyed individuals from the DHSS by using the National Insurance numbers and dates of birth, and then to trace them in the NHSCR in order to obtain information on follow-up status. After an initial pilot study at the DHSS on a sample from the computer file, it was decided that satisfactory individual linkage was possible. In addition, the DHSS was able to tell us whether, according to their records, each person was still alive or not. On the basis of this information, it was decided to follow up initially through the NHSCR only those reported as dead by the DHSS in order to obtain both death certificates and information on all of the areas in which they had been registered with a general practitioner during and since the original survey.

The methods used in the selection of the particular potteries and employees included in the survey are described in the HSE report (Fox et al., 1975). Briefly, the pottery industry was stratified into sectors according to the main product and the type of body used. A sample of the potteries in each stratum was taken and all workers in a selected pottery were invited to take part. A 94% participation rate was achieved. Valid records on 6187 persons from 40 potteries were identified from the available computer tape and these individuals are included in the mortality follow-up study.

Deaths among male workers between the date of the survey in each pottery and 31 July 1985 (tracing at the DHSS was completed between August 1985 and January 1986) have been compared with the numbers expected from the calendar-year, cause-, sex- and age-specific death rates for England and Wales, using the person-years method of analysis.

In addition to this national comparison, an adjustment has been made to the expected numbers to take account of the mortality levels in the areas in which the potteries were located and the workers lived. To do this, the standardized mortality ratios (SMRs) by sex, cause and local authority area during 1968–78, available from geographical atlases of mortality for those years (Gardner et al., 1983, 1984), were used. The number of expected deaths at national rates in each pottery was multiplied by the local SMR, and the results summed over potteries to produce a locally adjusted expected number. These locally adjusted comparisons are regarded as more appropriate than those based on the national death rates.

In addition, for lung cancer, to take account of the relationship with cigarette smoking, an adjustment was also made to the nationally calculated expected numbers by multiplying them in each smoking category (non-smokers, ex-smokers and current smokers at the time of the cross-sectional survey) by 0.10, 0.44 and 1.34 respectively, and then summing the resulting figures over smoking categories to produce an expected number adjusted for smoking habits. The multipliers used are estimates for men derived by Berry et al. (1985). Again, these smoking-habit-adjusted comparisons are regarded as more appropriate than those based on the national lung cancer death rate.

Confidence intervals for the ratios of observed to expected deaths and one-sided p values have been calculated by standard methods based on the Poisson distribution.

Results

Study sample

The number of potteries in the sample and the number of pottery workers included by product group and sex are shown in Table 1, these figures differing only slightly from those given in the HSE report (Fox et al., 1975). The potteries were located predominantly in the Midlands region of the country – 26 out of the total of 40 – but also in the north-west, London and other areas, including Scotland. The sample of workers surveyed included 4093 (66%) men as compared

Table 1. Number of potteries in the sample and number of pottery workers by product group and sex

Product group	Number of potteries	No. of workers		
		Men	Women	Total
Sanitary ware - vitreous china	6	1494	176	1670
Sanitary ware - fireclay	5	334	64	398
Domestic ware - earthenware, etc.[a]	5	434	485	919
Domestic ware - bone china	4	164	383	547
Wall tiles	2	475	402	877
Floor tiles	3	99	48	147
Industrial - earthenware, etc.[a]	3	355	318	673
Industrial - fireclay	5	423	112	535
Industrial - other	2	143	6	149
Body materials	3	97	5	102
Glazes, colours and transfers	2	75	95	170
Total	40	4093	2094	6187

[a] Earthenware, vitreous china and stoneware

with 2094 (34%) women. The largest product group in the survey was 'sanitary ware – vitreous china', with 1670 workers.

Table 2 shows the number of men and women in the study by job group. The largest categories are primary shaping, with 1201 (19%) workers, firing and warehousing, with 1147 (19%) workers, and other processes with 1379 (22%) workers, including fitters, cleaners, packers, etc.

Table 3 gives the age and sex distribution of the pottery workers at the time of the cross-sectional survey during 1970 and 1971. There is a fairly uniform distribution of ages over the working range with more than 1000 in each ten-year age group.

Follow-up: pilot study

Identification particulars (National Insurance number and date of birth) on a systematic sample of 1 in 30 of the workers from the computer file, i.e., 200 individuals, to cover the range of potteries, were sent to the DHSS, the results of whose search for names and vital status are shown in Table 4, together with the

Table 2. Number of pottery workers by job group and sex

Job group	No. of workers		
	Men	Women	Total
Material preparation	153	2	155
Body preparation	177	1	178
Primary shaping	869	332	1201
Secondary shaping	108	204	312
Other clay-shop workers	347	96	443
Firing and warehousing	783	364	1147
Glazing and decorating	254	575	829
Other processes	1145	234	1379
Office workers	257	286	543
Total	4093	2094	6187

Table 3. Number of pottery workers by age at the time of the survey in 1970-71 and sex

Age group (years)	No. of workers		
	Men	Women	Total
< 20	292	304	596
20-29	842	394	1236
30-39	808	339	1147
40-49	863	495	1358
50-59	864	468	1332
60+	424	94	518
Total	4093	2094	6187

Table 4. Number of individual records in the pilot study sent to the Department of Health and Social Security (DHSS) and the National Health Service Central Register (NHSCR) by outcome

NHSCR outcome	DHSS outcome			Total
	Alive	Dead	No trace	
Alive	143	-	-	143
Dead	4	30	-	34
Emigrated	1	-	-	1
No trace	12	-	-	12
Not sent	-	-	10	10
Total	160	30	10	200

outcome of the subsequent NHSCR search on the 190 who were traced at the DHSS. Of the sample of 200, 160 (80%) were reported as alive, 30 (15%) as dead, while for the remaining ten (5%) no trace of a person matching the particulars forwarded was found. At the NHSCR, all the 30 deaths reported by the DHSS were confirmed. Of the 160 persons reported as alive by the DHSS, and whose names were provided, 143 (89%) were reported likewise to be alive by the NHSCR. Of the remaining 17, four (2.5%) were reported as dead, one (0.6%) as emigrated and 12 (7.5%) could not be traced.

The four deaths included two with mention of the occupation 'pottery worker' on their death certificates (suggesting that the DHSS linkage was correct but that the vital status reported by the DHSS was wrong); for one, the occupation 'retired fork-lift truck driver' had been recorded. The other death was dated March 1969, which was earlier than the survey period; this suggests that the DHSS and/or NHSCR linkage was wrong. The dates of the other three deaths were not particularly recent – in 1980, 1982 and 1985 (January) – and they should have been notified to the DHSS by the time that they carried out their search. Among the 12 not traced by the NHSCR, there were three where the sex reported by the DHSS was different to that on the computer file – while for the remaining nine there was no apparent reason for non-tracing. Of the 143 reported alive by the NHSCR, 140 had been registered with general practitioners in the appropriate area for the pottery in which they were working at the time of the survey, two were registered outside their pottery area, and for one the registration is at present unknown.

We concluded from this pilot study that the follow-up information from the DHSS was sufficiently good to form the basis of a preliminary analysis of mortality among the workers.

Follow-up: main study

Following the pilot study, the identification particulars (National Insurance number and date of birth) from the remaining computer records were sent to the DHSS. The outcome, by age at survey, for all 6187 individuals at the DHSS, including those in the pilot study, is shown in Table 5. Overall, 7% (422) were not traced from the information supplied, and this percentage was the same for each sex and similar to that in the pilot study. Of the men, 78% (3205) were recorded

Table 5. Number (and percentage) of individual records sent to the Department of Health and Social Security (DHSS) by age at the survey in 1970-71 and outcome

Age group (years)	DHSS outcome						Total	
	Alive		Dead		No trace			
	No.	%	No.	%	No.	%	No.	%
<20	564	95	3	1	29	5	596	100
20-29	1168	94	8	1	60	5	1236	100
30-39	1042	91	38	3	67	6	1147	100
40-9	1116	82	162	12	80	6	1358	100
50-59	847	64	357	27	128	10	1332	100
60+	281	54	179	35	58	11	518	100
Total	5018	81	747	12	422	7	6187	100

as alive and 15% (615) as dead, as compared with figures of 87% (1813) and 6% (132) respectively for the women; the higher proportion of deaths among the men is what would be expected from the male/female differential in mortality rates and the age distributions by sex in 1970-71 (see Table 3). When looked at by age, the expected trends in proportions of alive and dead are seen, but also some indication that the percentage of persons not traced at the DHSS was higher among the older age groups. There were no important differences in the DHSS trace rates as between the 40 potteries.

During the tracing, the linkage between a pottery worker computer record and a DHSS record was not always found to be exactly in accordance with the computer information forwarded. Unamended linkage was achieved for 90% (5183) of traced individuals. Of the 582 individuals for whom changes were made at the DHSS to the information supplied, 61 individuals had amended National Insurance numbers and 553 had amended dates of birth. In addition, for 78 of the reported linkages the person was of a different sex to that indicated on the computer record.

Details of 743 of the 747 deaths reported by the DHSS, including the 30 from the pilot study, were then sent to the NHSCR with requests for death certificates including cause of death. For four of the reported deaths, details were not sent to the NHSCR because of a sex difference between the computer file and the DHSS return. On 445 (60%) of the 739 death certificates received (one was duplicated, three are still being traced), the occupation of the deceased was recorded in terms descriptive of a 'pottery worker' and 52 (7%) women were recorded as 'wife of' or 'widow of' their husband's occupation. Of the 242 (33%) with other jobs, 120 (16%) were consistent with other non-pottery employment from their job history on the computer record. For only three (0.4%) of the 739 did the general practitioner registration particulars indicate that they had lived in an area such that it was unlikely for them to have been pottery workers. These three individuals were amongst those for whom changes in date of birth had had to be made at the DHSS in order to achieve a linkage.

Follow-up analysis

The first person-years analyses carried out were to examine whether or not the results shown in Table 5 were indicative of a loss to tracing of older persons who had in fact died. For this purpose, workers who were not traced at the DHSS, those for whom changes in their particulars were needed to achieve a linkage, and the three individuals still being traced at the NHSCR were omitted. Table 6 shows the results for men and women together for all causes of death by calendar period and for: (a) age at survey; and (b) age during follow-up. In each section of the Table an indication is given of deficits in observed mortality rates among older workers and in the early calendar years after the survey. These figures suggested a possible problem in notification of vital status of persons from the DHSS around and after retirement age and particularly, though not only, in the early 1970s. This has, in fact, proved to be true, and to be related to a change from a manual to a computer system at the DHSS. Because of this problem, and because the number of expected deaths among women is much smaller, the preliminary results on mortality in this report are limited to men under 60 years of age at their survey date. Meanwhile more extensive tracing is being undertaken for the older workers.

Table 6. **Observed (O) and expected (E) mortality from all causes among pottery workers by calendar period and age**

Age (years)	Calendar period									Total		
	1970-75			1976-80			1981-85					
	O	E	O/E	O	E	O/E	O	E	O/E	O	E	O/E
Age at survey in 1970-71												
Under 60	96	118	0.8	171	164	1.0	218	196	1.1	485	478	1.0
60+	23	73	0.3	65	96	0.7	68	102	0.7	156	271	0.6
Total	119	192	0.6	236	260	0.9	286	298	1.0	641	750	0.9
Age during follow-up												
Under 60	76	89	0.9	89	78	1.1	53	59	0.9	218	226	1.0
60+	43	103	0.4	147	182	0.8	233	239	1.0	423	524	0.8
Total	119	192	0.6	236	260	0.9	286	298	1.0	641	750	0.9

Table 7 shows observed and expected mortality among men aged under 60 years at the survey by cause of death. Overall, as compared with rates for England and Wales, there has been a reported 7% excess death rate. For cancer mortality, there has been a 15% excess, largely attributable to the 40% excess in lung cancer and 60% excess in stomach cancer. The rates for circulatory and respiratory diseases are also somewhat in excess, but not for other causes of death - in particular for deaths certified as due to injury and poisoning, for which there is a notable deficit.

Table 7. Observed (O) and expected (E) mortality up to 31 July for cancers and other main causes of death among men under age 60 years at the survey in 1970-71: national and local comparisons

Cause of death		National rates		Locally adjusted rates	
	O	E	O/E	E	O/E
Malignant neoplasms:	118	102.6	1.15	107.6	1.10
Oesophagus	4	3.4	1.18	3.8	1.05
Stomach	15	9.4	1.60	11.9	1.26
Large intestine	3	6.1	0.50	6.0	0.50
Rectum	8	4.5	1.78	4.6	1.75
Pancreas	3	4.6	0.65	4.9	0.61
Lung	60	42.8	1.40[a]	45.6	1.32[b]
Prostate	3	3.6	0.84	2.8	1.07
Bladder	3	3.4	0.89	3.3	0.92
Brain	0	3.2	0	2.9	0
Other cancers	19	21.6	0.88	21.8	0.87
Circulatory disease	205	179.5	1.14	196.3	1.04
Respiratory disease	39	32.4	1.20	38.5	1.01
Digestive disease	7	9.3	0.75	9.7	0.72
Injury and poisoning	5	21.5	0.23	22.2	0.23
Other causes	16	18.1	0.88	20.4	0.78
All causes	390	363.4	1.07	394.7	0.99

[a] 95% confidence interval 1.07-1.80; $p=0.007$
[b] 95% confidence interval 1.00-1.69; $p=0.023$

The locally adjusted comparisons, also given in Table 7, show a somewhat different pattern, many ratios of observed/expected mortality being lower than in the national comparisons; this is because of the higher general population death rates in the mainly urban areas in which the potteries are located. Levels of overall mortality, as well as of mortality due to circulatory and respiratory disease, are similar to expected. However, an excess of lung cancer remains and, based on smaller numbers, of stomach and rectal cancers. Deaths from injury and poisoning still show the remarkable deficit.

Table 8 shows observed and expected mortality by number of years since first employment in the pottery industry for the main causes of death, using locally adjusted comparisons. No strong trends with time are found, although there is a general increase in observed/expected ratios, except for stomach cancer, other cancers and other causes. There is little indication in the Table of the healthy worker effect commonly seen in occupational groups, although the observed/expected (O/E) ratio within ten years of entering the industry is 0.7 (O=18, E=25.0).

Among men under 60 years of age at the time of the survey in 1970-71, 16% of the pottery workers were non-smokers, 15% ex-smokers and 68% current smokers of cigarettes. Table 9 shows the number of observed and expected deaths from lung cancer and other main causes by cigarette smoking status. A strong trend in the observed/expected ratios for lung cancer is found, but not for other cancers combined. There are also upward trends with cigarette smoking for circulatory and respiratory diseases. These figures confirm that sensible record linkage has been achieved.

Table 8. Observed (O) and expected (E) mortality by main causes of death up to 31 July 1985 among men under age 60 years at the time of the survey in 1970-71 by time since first employment in the pottery: locally adjusted comparisons

Cause of death	Years since entering employment								
	0-19			20-39			40+		
	O	E	O/E	O	E	O/E	O	E	O/E
Lung cancer	10	9.4	1.1	21	15.9	1.3	29	20.4	1.4
Stomach cancer	6	2.3	2.6	2	4.2	0.5	7	5.4	1.3
Other cancer	13	12.1	1.1	10	18.2	0.5	20	19.8	1.0
Circulatory disease	40	42.2	0.9	77	71.5	1.1	88	82.6	1.1
Respiratory disease	6	7.9	0.8	14	12.4	1.1	19	18.3	1.0
Other causes	11	18.9	0.6	9	19.2	0.5	8	14.2	0.6
All causes	86	92.8	0.9	133	141.3	0.9	171	160.6	1.1

Table 9. Observed (O) and expected (E) mortality by main causes of death up to 31 July 1985 among men under age 60 years at the time of the survey in 1970-71 by cigarette smoking habits as reported at the time: national comparisons and adjustment for smoking category for lung cancer

Cause of death	Type of smoker								
	Non-smoker			Current smoker			Ex-smoker		
	O	E	O/E	O	E	O/E	O	E	O/E
Lung cancer:									
Adjusted[a]	0	0.4	0	3	3.9	0.8	57	40.3	1.4
Unadjusted[a]	0	3.9	0	3	8.9	0.3	57	30.1	1.9
Other cancer	6	6.0	1.0	15	11.9	1.3	37	41.8	0.9
Circulatory disease	3	17.0	0.2	38	36.5	1.0	164	125.9	1.3
Respiratory disease	2	3.0	0.7	6	6.8	0.9	31	22.6	1.4
All causes	15	36.5	0.4	67	72.7	0.9	308	254.1	1.2

[a] For the cigarette smoking habits of the pottery workers

When adjustments are made to the nationally expected numbers of lung cancer deaths by smoking category, the adjusted expected numbers are closer to those observed. However, there is still an excess of lung cancer mortality among current smokers, and overall there were 60 observed deaths as compared with a smoking-adjusted expected figure of 44.6, giving an observed/expected ratio of 1.34 (95% confidence interval 1.03-1.73; $p=0.016$). It is of interest to note how closely this corresponds to the locally adjusted ratio for lung cancer (1.32) in Table 7, where the purpose of the local adjustment is mainly to allow for the well-known higher smoking levels in urban as compared with rural areas.

Table 10 shows the number of observed and expected lung cancer deaths among men under 60 years of age at the survey by: (a) product group at survey; and (b) job group at survey, using both national and smoking-adjusted comparisons. In terms of specific products and job groups, there is a range of deficits and excesses, most based on small numbers, with one job group (glazing and decorating) having a statistically significant excess in both comparisons.

Table 10. Observed (O) and expected (E) lung cancer mortality up to 31 July 1985 among men under age 60 years at the time of the survey in 1970–71 by product group and job group: national and smoking-adjusted comparisons

Product group		National rates			Smoking-adjusted rates	
	O	E	O/E		E	O/E
Product group						
Sanitary ware – vitreous china	16	15.5	1.03		16.5	0.97
Sanitary ware – fireclay	7	4.2	1.68		4.3	1.62
Domestic ware – earthenware, etc.[a]	6	3.5	1.69		3.5	1.71
Domestic ware – bone china	1	1.3	0.74		1.4	0.70
Wall tiles	8	5.3	1.50		5.5	1.45
Floor tiles	2	1.7	1.17		1.6	1.24
Industrial – earthenware, etc.[a]	6	4.0	1.51		3.9	1.53
Industrial – fireclay	8	3.8	2.10		4.2	1.92
Industrial – other	3	1.3	2.36		1.2	2.44
Body materials	3	1.4	2.18		1.7	1.76
Glazes, colours and transfers	0	0.8	0		0.7	0
Total	60	42.8	1.40		44.6	1.34
Job group						
Material preparation	5	1.8	2.77		2.1	2.34
Body preparation	3	1.9	1.59		2.0	1.52
Primary shaping	9	7.9	1.14		8.1	1.11
Secondary shaping	0	0.8	0		0.8	0
Other clay-shop workers	8	3.7	2.19		4.2	1.89
Firing and warehousing	12	9.3	1.29		9.6	1.26
Glazing and decorating	6	2.0	3.01[b]		2.1	2.82[c]
Other processes	16	12.9	1.25		13.1	1.22
Office workers	1	2.6	0.38		2.5	0.39
Total	60	42.8	1.40		44.6	1.34

[a] Earthenware, vitreous china and stoneware
[b] 95% confidence interval 1.11–6.56; $p = 0.017$
[c] 95% confidence interval 1.03–6.13; $p = 0.020$

Table 11 shows the results for lung cancer in relation to available information on dust exposure. Mean respirable quartz concentrations in the workplace at the time of the original survey were available for most job/product group combinations, and varied from under 0.01 mg/m^3 for glazing and decorating in the bone china sector to 0.20 mg/m^3 for primary shaping in the industrial (other) product group (Fox et al., 1975). Using these values, the men were combined so as to form four groups of similar size, firstly by exposure level at the time of the survey and secondly by cumulative exposure on the assumption that current dust concen-

Table 11. Observed (O) and expected (E) lung cancer mortality up to 31 July 1985 in relation to respirable quartz exposure at the time of the survey in 1970–71 and cumulative exposure up to 1970–71: national and smoking-adjusted comparisons

Exposure category (mg/m^3)	O	National rates		Smoking-adjusted rates	
		E	O/E	E	O/E
Respirable quartz concentration at the time of the survey					
0–0.02	15	9.0	1.67	8.9	1.68
0.03–0.04	14	12.4	1.13	13.2	1.06
0.05–0.09	13	8.5	1.53	9.1	1.43
0.1+	17	9.8	1.73[a]	10.5	1.63
Cumulative exposure to respirable quartz up to the time of the survey					
0–0.14	5	4.2	1.18	4.6	1.08
0.15–0.49	8	7.9	1.01	8.1	0.99
0.50–1.49	25	14.8	1.68[b]	15.4	1.62[c]
1.50+	21	13.3	1.58	13.9	1.51

[a] 95% confidence interval 1.01–2.78; $p=0.023$
[b] 95% confidence interval 1.0–2.49; $p=0.010$
[c] 95% confidence interval 1.05–2.39; $p=0.015$

trations applied to the entire occupational history in the potteries. In each of these analyses, men in office work and the glazes product group were excluded because of the lack of hygiene measurements. The lung cancer relative risks are seen to be unrelated to current exposure at the time of the survey, but there are apparent excesses in each of the two higher groups by cumulative exposure but not in the two lower groups, using both national and smoking-adjusted comparisons.

Discussion

This paper reports the preliminary results of a follow-up study of pottery workers over a 15-year period. Because of the destruction of the original questionnaire records on the workers, but not the computer tape abstracts, the main concern to date has been to identify, activate and evaluate a method of follow-up that would not normally be used. This has been judged to have produced appropriate data, at least on the subset of workers who were under the age of 60 years at the time of the original respiratory disease survey in 1970–71. However, some small doubts remain and more detailed investigation of the tracing mechanism is under way, including workers over the age of 60 years who have been omitted from the present analysis because of concern about bias in the follow-up.

These initial results show that, among men under 60 years of age at the time of the original survey, overall mortality is similar to that expected both nationally and locally. There is only weak evidence of any healthy worker effect, which may relate to the cross-sectional manner in which the study cohort was assembled. The findings indicate an excess of lung cancer mortality, with the expected strong relationship with the cigarette smoking habits of the workers. Adjustment for these habits, however, does not explain the excess. No particular excesses by product or job group were found, but there was an indication of a relationship with estimated cumulative exposure to respiratory quartz during employment in the potteries. Interpretation of these findings on lung cancer is uncertain at present, but there is certainly a suggestion of an association with working in the pottery industries.

There is also an excess of stomach cancer in this preliminary analysis, but this largely disappears after local area rates are taken into account, and an excess, based on small numbers, of rectal cancer. No excess of non-malignant respiratory diseases has been found, and only one death from silicosis has been reported. There is a substantial reported deficit of deaths from injury and poisoning in both sexes, for which we have no explanation at present.

After the tracing rate among the older workers has been improved as far as possible, it is intended to analyse other factors measured in the original survey. These will include lung function measures (forced expiratory volume in one second and forced vital capacity), respiratory symptomatology (from the Medical Research Council questionnaire), radiographic findings (using ILO categories) and previous occupational history.

Acknowledgements

We are grateful to the Department of Health and Social Security and the National Health Service Central Register for their help in tracing records which made this study possible. We should like to thank Mrs Brigid Howells for typing the manuscript. Financial support for the study was provided by the International Agency for Research on Cancer, Lyon, and Dr A.C. Fletcher was supported by the Health and Safety Executive.

References

Berry, G., Newhouse, M.L. & Antonis, P. (1985) Combined effect of asbestos and smoking on mortality from lung cancer and mesothelioma in factory workers. *Br. J. Ind. Med.*, 42, 12-18

Fox, A.J., Greenberg, M. & Ritchie, G.L. (1975) *A Survey of Respiratory Disease in the Pottery Industry*, London, HMSO

Gardner, M.J., Winter, P.D., Taylor, C.P. & Acheson, E.D. (1983) *Atlas of Cancer Mortality in England and Wales, 1968-78*, Chichester, Wiley

Gardner, M.J., Winter, P.D. & Barker, D.J.P. (1984) *Atlas of Mortality from Selected Diseases in England and Wales, 1968-78*, Chichester, Wiley

IARC (1987) *IARC Monographs on the Evaluation of the Carcinogenic Risk of Chemicals to Humans, Vol. 42, Silica and Some Silicates*, Lyon, International Agency for Research on Cancer

Lung cancer risk among pneumoconiosis patients in Japan, with special reference to silicotics

K. Chiyotani[1], K. Saito[1], T. Okubo[2] and K. Takahashi[2]

[1]Rosai Hospital for Silicosis, Tochigi Prefecture, Japan
[2]University of Occupational and Environmental Health,
Kitakyushu City, Japan

Summary. A cohort study was conducted on 3335 hospitalized male pneumoconiosis patients in order to determine their lung cancer mortality. The patients were collected from 11 hospitals specializing in the treatment of pneumoconiosis and other occupational diseases and injuries, and were followed up during 1979–1983. Age-adjusted expected numbers were calculated using the mortality of the general population as a standard. A significantly high O/E ratio (4.80) for lung cancer was observed, particularly among silicotics (6.03). A slightly more than two-fold increase in risk for lung cancer was found among those who had never smoked. No dose–response relationship was found between lung cancer mortality and duration of dusty work or radiographic categories. A nested case–control analysis showed work experience in the ceramics industry to be a risk factor for lung cancer.

Introduction

Although a well designed epidemiological study on lung cancer risk among workers exposed to silica dust has not yet been conducted in Japan, it has been reported, based both on patients and on autopsy material, that the lung cancer risk is higher in pneumoconiosis patients than in the general population (Chiyotani, 1983). From the earlier reports, however, it was not clear whether this higher risk was restricted to specific hospitals or was common to all.

A joint survey project by 11 'Rosai' hospitals was organized in 1979 to clarify the matter. The 'Rosai' hospital system is operated by the Japanese Workmen's Compensation Insurance, and comprises 37 hospitals throughout Japan. The 11 hospitals participating in the present study are major hospitals for the treatment of pneumoconiosis in the 'Rosai' system. Although there is no obligation for pneumoconiosis patients to be sent to these hospitals, the system has played an important part in the treatment of the disease. It is estimated that approximately 8.7% of all pneumoconiosis patients in Japan (of whom there are currently some 35 000) receiving compensation from the insurance system are being treated in 'Rosai' hospitals, 73% of them in the 11 participating hospitals.

Pneumoconiosis patients in the participating hospitals at the beginning of 1979 and new patients thereafter were followed up for five years, i.e., until the end of

1983, in order to determine the frequency of deaths from lung cancer. Cases in which death was due to lung cancer satisfying the diagnostic definition adopted were then used for a nested case–control study and exposure history was analysed in detail.

Materials

Patients diagnosed as suffering from pneumoconiosis and qualifying for workmen's compensation before or during the study period in one of the 11 participating hospitals were included in the study. Information on each patient regarding work history, smoking and medical examinations were collected, together with demographic data. The number of female patients was too small for analysis and they were therefore excluded from the study. When a patient on the list died during the study period, each doctor in charge of the patient was requested to report the major cause of death based on both the clinical and histopathological diagnosis.

The number of patients included in this study is shown by type of pneumoconiosis in Table 1. There were few cases of types other than silicosis and anthracosilicosis.

Table 1. Number and percentage distribution of subjects by type of pneumoconiosis

Type of pneumoconiosis	Number	%
Silicosis	1941	58.2
Anthracosilicosis	1278	38.3
Asbestosis	22	0.7
Caused by other silicates	37	1.1
Caused by other inorganic dusts	57	1.7
Total	3335	100.0

Table 2 shows the distribution of these patients by time of initial diagnosis and age at that time. Of the total, 36% joined the group after the beginning of the study, while the remainder had already been diagnosed as suffering from pneumoconiosis and treated in the hospitals. The earliest diagnosis was made in 1950 and a substantial proportion (696 cases, or 20.9%) had been undergoing treatment for more than ten years.

Method

Follow-up

The study population was followed up from 1 January 1979 to 31 December 1983. The first year of follow-up for each patient was excluded from the calculation of person-years, and deaths occurring within that period were not included in the number of deaths. Although this procedure resulted in a substantial decrease in the number observed, it was considered to be essential in order to avoid selection bias due to the tendency for patients already diagnosed, and especially those with lung cancer, to be referred to highly specialized hospitals, such as the 'Rosai' hospitals.

Table 2. Distribution of subjects by date of initial diagnosis and age in ten-year intervals

Date first diagnosed	Age at initial diagnosis (years)					
	20-39	40-49	50-59	60-69	70+	Total
1954	5	11	2	0	0	18
1955-1959	21	63	44	0	0	128
1960-1964	29	83	106	17	0	235
1965-1969	24	60	154	70	7	315
1970-1974	16	115	219	184	38	572
1975-1978	14	122	308	319	92	855
1979-	4	139	450	417	202	1212
Total	113	593	1283	1007	339	3335

Person-years and standardized mortality ratios

Person-years after one year from initial diagnosis were counted by calendar year five-year age class. The age-specific mortality rate for major causes of death among the general Japanese male population in 1982 was used for the calculation of the age-adjusted expected number of deaths, and the standardized mortality ratios (SMRs) were calculated. In this calculation, the clinical diagnosis was used in the classification of the cause of death so that the accuracy of the diagnosis would be closer to that of the vital statistics. Statistical significance was expressed in terms of the 95% confidence interval.

Case-control study

A case-control study was conducted for 73 lung cancer cases selected from those observed in the follow-up study by applying more strict diagnostic criteria than those used in that study; lung cancer had to be confirmed by pathological diagnosis at autopsy and/or biopsy. The work and exposure history of each case was compared to that of a matched control, selected from the pneumoconiosis patients in the follow-up study. The matching criteria used for selecting controls were as follows:

(1) Pneumoconiosis patient in the same hospital.
(2) Alive at the time of death of the subject in the case-control study.
(3) Within ± three years of the subject's age at the time.
(4) In the same smoking category (non-smoker versus ex- or current smoker).

A control case meeting the above criteria could not be found for one lung cancer case. As a result, 72 pairs were used for the study.

Results

Observed (O) and expected (E) numbers of death from major causes during the study period are shown in Table 3 together with the 95% confidence intervals.

Table 3. Mortality from selected causes[a]

Cause of death[b]	O	E	O/E	95% CI
All causes	581	205.4	2.83	2.69-2.97
Neoplasms (140-208)	129	63.0	2.05	1.80-2.30
Oesophagus (150)	7	3.1	2.26	1.12-3.39
Stomach (151)	26	19.3	1.35	0.89-1.80
Liver (155)	6	7.0	0.86	0.10-1.61
Pancreas (157)	7	3.4	2.06	0.97-3.14
Lung (162)	60	12.5	4.80	4.23-5.37
Leukaemia (204-208)	2	1.1	1.82	0.00-3.73
Ischaemic heart diseases (410-414)	8	16.1	0.50	0.00-1.00
Cerebrovascular diseases (430-438)	28	42.2	0.66	0.36-0.97
Chronic liver diseases and liver cirrhosis (571)	11	5.4	2.04	1.18-2.90

[a] O, observed number of deaths; E, expected number of deaths; 95% CI, 95% confidence interval
[b] Item numbers from the Ninth Revision of the *International Classification of Diseases* in parentheses

In addition to the data shown in the table, there were 345 deaths from respiratory insufficiency due to pneumoconiosis. However, since an appropriate standardized mortality ratio for this cause of death was not available for use in calculating the expected number, it was not included in the table. The O/E ratio for lung cancer was significantly higher than that for other causes. The figures for the various hospitals differed only slightly from one another.

Smoking status at first diagnosis is routinely recorded at the member hospitals. Table 4 shows the O/E ratios by smoking status for lung and stomach cancer. Smoking information was not available for a very small number of individuals (eight, or 10.6 person-years), who were not included in the Table.

Table 4. Observed (O) and expected (E) death rates for lung cancer (ICD 162) and stomach cancer (ICD 151) in relation to smoking category

Smoking category	Person-years	Cause of death							
		Lung cancer				Stomach cancer			
		O	E	O/E	95% CI[a]	O	E	O/E	95% CI[a]
Never smoked	1229.3	4	1.8	2.22	0.73-3.71	3	2.6	1.15	0.00-2.39
Ex-smoker	3883.0	23	4.6	5.00	4.07-5.93	11	7.1	1.55	0.80-2.30
Current smoker	5894.5	33	6.1	5.41	4.60-6.22	12	9.5	1.26	0.61-1.91

[a] 95% confidence interval

Table 5 shows the O/E ratio for lung cancer by duration of exposure to dust. This was reported for each pneumoconiosis patient by place of exposure, i.e., underground or surface. The duration shown is the total for both locations.

Table 5. O/E ratio for deceased cases of lung cancer (ICD 162) in relation to length of exposure[a]

Length of exposure (years)	Person-years	O	E	O/E	95% CI
< 19	4022.6	15	3.6	4.17	3.11-5.22
20-29	4221.3	26	4.6	5.65	4.72-6.58
30 <	2773.4	19	4.3	4.42	3.45-5.38

[a] O, observed number of deaths; E, expected number of deaths; 95% CI, 95% confidence interval

O/E ratios by type of pneumoconiosis are shown in Table 6. Since the numbers of cases of types other than silicosis and anthracosilicosis were small, this almost amounts to a comparison between these two types. The frequency of lung cancer among silicotics was twice that among anthracosilicosis patients.

The frequency of lung cancer by radiographic category of pneumoconiosis at the initial diagnosis is shown in Table 7. In Japan, the chest X-ray film of a newly diagnosed pneumoconiosis patient must be submitted to a committee of the Ministry of Labour. Qualified specialists on this committee decide as to the radiographic category and whether the patient should receive compensation under the Pneumoconiosis Law or not and to which health administrative category he should be assigned. The classification used for the purposes of this analysis is the Japanese classification, which is almost the same in categories I–III as the ILO classification, but includes a category IV for large opacities. Higher O/E ratios were observed in the lower radiographic categories.

Table 8 summarizes the results of the nested case–control study. With the classification of the histological types of lung cancer used in this study, epidermoid carcinoma accounted for 57.5%, small-cell carcinoma for 21.8%, adenocarcinoma for 11.5% and large-cell carcinoma for 9.2% of cases. In the case–control study, the comparisons were repeated separately for the epidermoid cancer subgroup. The odds ratio for silicosis was significantly elevated and, for the epidermoid subgroup, was twice that for the total group. For the various work histories, significantly higher odds ratios were observed only for kilnmen or casters and for the ceramics industry. All six pairs of 'case-yes', 'control-no' for these two types of exposure were in the epidermoid subgroup (Table 9).

Discussion

The subjects in the present study were pneumoconiosis patients who had survived for various periods of time since the initial diagnosis together with patients newly diagnosed during the study period. Because of the nature of the study design, some selection bias was unavoidable. No information is available on patient

Table 6. Mortality from selected causes by type of pneumoconiosis[a]

Cause of death[b]	Type of pneumoconiosis											
	Silicosis				Anthracosilicosis				Other			
	O	E	O/E	95% CI	O	E	O/E	95% CI	O	E	O/E	95% CI
All causes	352	120.2	2.93	2.75–3.11	210	80.4	2.61	2.39–2.83	19	4.8	3.96	3.05–4.87
Neoplasms (14–208)	86	37.2	2.31	1.98–2.64	40	24.2	1.65	1.25–2.06	3	1.5	2.00	0.37–3.63
Oesophagus (150)	4	1.9	2.11	0.65–3.56	3	1.2	2.50	0.67–4.33	0	0.1	0.00	–
Stomach (151)	14	11.4	1.23	0.64–1.82	11	7.4	1.49	0.75–2.22	1	0.4	2.50	–
Liver (155)	4	4.2	0.95	0.00–1.93	2	3.0	0.67	0.00–1.82	0	0.2	0.00	–
Pancreas (157)	6	2.0	3.00	1.59–4.41	1	1.3	0.77	0.00–2.52	0	0.1	0.00	–
Lung (162)	44	7.3	6.03	5.29–6.77	15	4.9	3.06	2.16–3.96	1	0.3	3.33	–
Leukaemia (204–208)	2	0.7	2.86	0.47–5.25	0	0.4	0.00	–	0	0.0	0.00	–
Ischaemic heart diseases (410–414)	3	9.4	0.32	0.00–0.97	4	6.3	0.63	0.00–1.43	1	0.4	2.55	0.00–5.66
Cerebrovascular diseases (430–438)	18	24.5	0.73	0.33–1.14	9	16.8	0.54	0.05–1.02	1	0.9	1.11	0.00–3.22
Chronic liver diseases and liver cirrhosis (571)	6	3.3	1.82	0.72–2.92	5	2.0	2.50	1.09–3.91	0	0.1	0.00	–

[a] O, observed number of deaths; E, expected number of deaths; 95% CI, 95% confidence interval; this was not calculated if E was less than 0.5
[b] Item numbers from the Ninth Revision of the *International Classification of Diseases* in parentheses

Table 7. O/E ratio for deceased cases of lung cancer (ICD 162) by radiographic category[a]

Radiographic category[b]	Person-years	O	E	O/E	95% CI
1	1713.9	11	1.8	6.11	4.63-7.60
2	4558.8	29	5.7	5.09	4.25-5.93
3	2137.2	9	2.2	4.09	2.74-5.44
4	2607.5	11	2.7	4.07	2.86-5.29

[a] O, observed number of deaths; E, expected number of deaths; 95% CI, 95% confidence interval
[b] According to the *Japanese Classification of Radiographs of Pneumoconiosis*, 1960

Table 8. Summary of matched case-control comparisons by exposure, work history and industry[a]

Exposure, work history, and industry	Total pairs ($n=72$)		Epidermoid cancer subgroup pairs ($n=38$)	
	Odds ratio	Chi square	Odds ratio	Chi square
Silicosis	5.67	9.80	12.00	9.31
Duration (years) of dusty work on the surface:				
≥ 10	0.67	0.40	1.00	0.00
≥ 20	1.00	0.00	2.00	0.33
≥ 30	0.75	0.14	0.50	0.33
Age (years) at first engagement in dusty work:				
< 20	1.73	2.13	1.83	1.47
< 30	1.90	2.79	1.50	0.60
< 40	2.00	0.33	1.00	0.00
Grade of pneumoconiosis by radiographic category:				
> 2	1.25	0.22	1.67	0.50
> 3	1.00	0.00	1.20	0.09
4	0.90	0.05	1.20	0.11
Work history:				
Quarrymen	1.33	0.57	1.83	1.47
Quarrymen or coal miners	0.57	1.64	0.50	1.33
Quarrymen, coal miners or other mine workers	2.00	1.00	3.00	1.00
Kilnmen or casters	∞	6.00	∞	3.00
Industry:				
Mining	1.83	1.47	1.67	0.50
Mining or tunnelling	1.75	0.82	1.50	0.20
Ceramics	∞	6.00	∞	3.20

[a] For details of exposures associated with an excess risk, see table 9

Table 9. Exposures associated with an excess risk

Exposure	Case-yes, control-no	Case-no, control-no	Case-yes, control-yes	Case-no, control-yes	Relative risk
Silicosis (all pairs)	17	18	34	3	5.67
Epidermoid cancer subgroup pairs	12	8	17	1	12.00
Kilnmen or casters (all pairs)	6	66	0	0	∞
Ceramics industry (all pairs)	6	64	2	0	∞

Table 10. O/E ratios by year of initial diagnosis and major cause of death

Cause of death[a]	Year of initial diagnosis					
	-1959	1960-64	1965-69	1970-74	1975-78	1979-83
All causes	3.3	2.9	2.5	3.1	2.5	3.2
Neoplasms (140-208)	2.6	1.5	1.9	1.9	2.1	2.6
Stomach (151)	1.4	1.3	0.7	1.8	1.5	0.9
Lung (162)	6.2	3.9	4.2	3.8	5.1	6.9
Ischaemic heart diseases (410-438)	1.6	0.5	0.0	0.5	0.4	0.5
Cerebrovascular diseases (43-438)	1.2	0.9	1.0	0.2	0.5	0.8
Chronic liver diseases and liver cirrhosis (571)	5.8	1.7	2.7	2.4	0.0	4.2

[a]Item numbers from the Ninth Revision of the *International Classification of Diseases* in parentheses

mortality or morbidity during the period from the initial diagnosis until the beginning of the study in January 1979. However, as shown in Table 10, no large differences were seen in the O/E ratios for lung cancer or for other causes of death as between subgroups for which the time since initial diagnosis differed. This suggests that there was no large selection bias in the frequency of cause-specific mortality during the period in question.

Another possible source of bias may be the fact that the population of this study consists of patients in a large hospital system in Japan. Although there was not enough evidence to justify this, the first year following the diagnosis of pneumoconiosis was excluded in the calculation of both mortality and person-years. This was done to avoid the inclusion of cases of already diagnosed serious diseases, such as cancer, in the cohort. In fact, lung cancer was the only cause of death for which the mortality was clearly elevated and the mortality from all other major causes of death did not differ markedly from that of the general population. The elevated lung cancer mortality among pneumoconiosis patients cannot be explained simply by selection bias, and some other factor is believed to be associated with this phenomenon.

Smoking may play a major part in the increase in lung cancer deaths found in the present study. The proportion of current smokers and ex-smokers in the cohort was 10% higher than in the general population. However, the more than four-fold increase in lung cancer in this cohort cannot be completely explained by this high smoking rate. When the lung cancer mortality among the general population, which includes about 60% current smokers as well as 10–20% ex-smokers, was applied to the members of the cohort who had never smoked, the O/E ratio still indicated a more than two-fold increase. Even though some misclassification in smoking status was suggested by the relatively small difference in lung cancer frequency among those who had never smoked as compared with ex-smokers or current smokers, the above evidence may indicate that some risk factor plays a part in the increase in lung cancer in the cohort in the present study.

The period of assignment to dusty work was used as an indicator of exposure, but no relationship was found when the cohort was divided into three subgroups based on duration of exposure. However, duration of exposure is not necessarily the same as magnitude of exposure, and the existence of a dose–response relationship cannot be totally excluded.

A comparison by radiographic type suggests that the milder pneumoconiosis cases have a higher lung cancer risk. In earlier studies, when patients suffered more serious pneumoconiosis and their average life expectancy was significantly shorter than that of the general population, only a very slightly increased lung cancer risk was demonstrated. A high lung cancer mortality was observed at the 'Rosai' hospitals only after 1975, when the average life expectancy of hospitalized pneumoconiosis patients reached the age of predilection of lung cancer (Chiyotani, 1981).

Silicosis patients showed a higher lung cancer risk than those with anthracosilicosis and other pneumoconioses. A significantly high odds ratio for silicosis was also found in the case–control studies, as well as an indication that some factors in the ceramics industry may be related to lung cancer. However, only a small part of the elevated risk of lung cancer among silicosis patients can be explained in this way and the reason for the remainder remains unclear.

Conclusions

A significantly high O/E ratio (4.80) for lung cancer was observed in the cohort of 3335 hospitalized pneumoconiosis patients from 11 participating hospitals located throughout Japan. Of the various types of pneumoconiosis, silicosis was associated with the highest O/E ratio for lung cancer (6.03). A slightly more than two-fold increase in the O/E ratio was found among those who had never smoked. Even though some possibility of misclassification in smoking status was suggested by the result, the excess risk of lung cancer observed in the cohort could not be explained only by smoking.

The excess risk was not related to length of exposure, but the recent increase in the longevity of pneumoconiosis patients may be a factor in the high O/E ratio, which was not seen in the past.

Finally, a nested case–control study showed a strong association of lung cancer with the ceramics industry. However, only a part of the elevated lung cancer risk of the total cohort could be explained by this factor, and the major reason for the excess risk is not clear.

References

Chiyotani, K. (1981) Clinical study on association of lung cancer with pneumoconiosis [in Japanese]. *Jap. J. Traumatol. Occup. Med.*, 29, 221-228

Chiyotani, K. (1983) Excess of lung cancer deaths in hospitalized pneumoconiotic patients. In: *Proceedings of the VIth International Pneumoconiosis Conference 1983, Bochum, Federal Republic of Germany, 20-23 September*, Vol. 1, Bochum, Bergbau-Berufsgenossenschaft, pp. 228-236

Mortality from specific causes among silicotic subjects: a historical prospective study

F. Merlo[1], M. Doria[1], L. Fontana[2], M. Ceppi[1], E. Chesi[3] and L. Santi[4]

[1]*Department of Epidemiology and Biostatistics,
Istituto Nazionale per la Ricerca sul Cancro,
Genoa, Italy*

[2]*Occupational Health Department,
Ospedali Civili San Martino,
Genoa, Italy*

[3]*Computing Centre, Istituto Nazionale per la
Ricerca sul Cancro, Genoa, Italy*

[4]*Istituto di Oncologia, Università degli Studi di Genova,
Genoa, Italy*

Summary. A historical mortality study was conducted among 520 silicotic subjects diagnosed at the Department of Occupational Health of the San Martino Hospital, Genoa, Italy, between 1961 and 1980. Vital status was ascertained as of 1 January 1982. Age-, sex- and calendar-year-adjusted standardized mortality ratios (SMRs) for specific causes were computed using Italian as well as Genoa County male population death rates.

The study shows statistically significant increased mortality from all deaths (SMR=2.92), all cancers (SMR=2.38), respiratory tract cancers (SMR=6.85), respiratory tract diseases (SMR=13.63), and from 'other diseases' (SMR=6.81). The excess mortality from respiratory tract diseases and from 'other diseases' are mainly attributable to silicosis and silicotuberculosis, respectively.

These findings confirm the existence of a causal association between silicosis and increased mortality from both malignant and non-malignant respiratory tract diseases. The high mortality from respiratory tract cancers was still present even after adjustment for smoking.

Introduction

The hypothesis that subjects occupationally exposed to silica dust have an increased risk of lung cancer has been widely debated in the scientific literature (Ziskind *et al.*, 1976; Goldsmith & Guidotti, 1982; IARC, 1987). Several experimental studies have shown that silica is capable of inducing cancerous lesions (Wagner et al., 1980; Holland et al., 1986; Saffiotti, 1986). The results of most of the recent epidemiological studies tend to show that a causal relationship does exist between exposure to mixed silica dust and the development of respiratory tract cancers both in exposed workers (Sherson & Iversen, 1986; Palmer & Scott, 1981; Thomas, 1982; Finkelstein et al., 1986), and in subjects receiving compensation for silicosis (Santi & Balestra, 1957; Kurppa et al., 1986; Westerholm et al., 1986; Zambon et al., 1986). The possibility that the association between silica and lung cancer in man is confounded by the presence in the working environment and in cigarette smoke of other well known or suspected carcinogens has been suggested by several researchers (Heppleston, 1985; Archer et al., 1986; Palmer & Scott, 1986; Goldsmith & Guidotti, 1986).

Materials and methods

The aim of the study was to evaluate the suggested association between silicosis and lung cancer mortality in a cohort of 520 silicotic subjects identified among patients who were hospitalized at the Department of Occupational Health of San Martino Hospital, Genoa, Italy, during the period between 1 January 1961 and 31 December 1981. All the silicotics included in the study were exposed to mixed silica dust in the same geographical area where, until the late 1950s, tunnelling, slate quarrying, sandblasting, and refractory brick production were quite common. The study subjects were collected from the admission–discharge records of the above-mentioned Department of San Martino Hospital. Only individuals diagnosed as silicotics were included in the study. Occupational history, smoking habits, and year of, and age at diagnosis of silicosis were collected from the hospital records for each person. The diagnosis of silicosis was based on the result of chest X-rays and lung function tests. A historical cohort mortality study was performed for the period from 1 January 1961 to 1 January 1982. Vital status was determined by means of the registries of deaths of the county of Genoa or other cities (for those subjects who died outside Genoa). Underlying causes of death were coded by a nosologist according to the Ninth Revision of the *International Classification of Diseases*. Expected numbers of deaths were estimated by applying age-, cause- and calendar-year-specific mortality rates for the male population of Genoa and the Italian male population to the person-years of observation computed as described by Hill (Hill, 1972). Standardized mortality ratios (SMRs) were computed in order to estimate the risks both for overall mortality and for specific causes of death. The 95% confidence intervals (95% CI) for the SMRs were computed in order to assess the statistical significance of the differences between the observed and expected numbers of deaths.

Results

The vital status of the group of silicotic subjects at the end of the study is shown in Table 1, while Table 2 shows the distribution of person-years, deaths from all causes, and from malignant and non-malignant respiratory tract diseases by age groups.

Table 1. Vital status of the cohort of silicotics at the date of termination of the study (31 December 1981)

Vital status	No.	%
Alive	243	46.73
Deceased	225	43.26
Lost to follow-up	52	10.01
Total	520	100

Table 2. Person-years of observation and deaths from selected causes by age group

Age group	Person-years	Cause of death		
		All causes	Respiratory diseases	Respiratory cancers
25–44	65	3	–	
45–54	396	22	11	5
55–64	863	69	25	10
65–74	708	94	45	11
75+	213	37	21	–
Total	2245	225	102	26

All subjects were first exposed to silica dust when they were less than 25 years old; the distribution of the whole cohort by age at first diagnosis of silicosis is shown in Table 3. Table 4 shows the overall and cause-specific SMRs when the mortality experience of the study group is compared with that of the Italian male population. Overall mortality is increased significantly (SMR=2.92, 95% CI=2.57–3.34), the increase being due to the excess mortality from respiratory diseases (SMR=13.63, 95% CI=11.1–16.5, respiratory tract cancers (SMR=6.85, 95% CI=4.47–10), and 'other diseases' (SMR=6.81, 95% CI=4.77–9.43). It should be noted that the excess mortality from non-malignant respiratory diseases is largely attributable to silicosis: 84 out of 102 deaths from respiratory diseases showed the typical sequence of events resulting from silicosis (i.e., cor pulmonale and chronic respiratory failure). The mortality excess from 'other diseases' (36 observed deaths *versus* 5.2 expected) is almost totally explained by the inclusion in this group of 33 (91.6%) deaths from silicotuberculosis (item 011.4 in the Ninth Revision of the *International Classification of Diseases*). The increased risk for digestive tract diseases is mainly due to alcohol-related liver damage. Mortality was higher than expected for all cancers (SMR=2.38, 95% CI=1.70–3.24), and for respiratory tract cancers (SMR=6.85, 95% CI=4.47–10).

Table 3. Distribution of silicotics by age at first diagnosis of silicosis

	Unknown	< 25	25-34	35-44	45-54	55-64	65+	Total
No.	71	9	10	60	137	133	100	520
%	13.6	2	2	11.5	26.3	25.6	19.2	100

Table 4. Mortality from selected causes among silicotics, 1961-82[a]

Cause of death	ICD Code	O	E	SMR	95% CI
Malignant neoplasms:	140-209	40	16.8	2.38	1.70-3.24
Buccal cavity and pharynx	140-149.9	1	0.6	1.76	0.02-9.78
Oesophagus and stomach	150-151.9	4	3.7	1.09	0.29-2.78
Intestine and rectum	152-154.9	2	1.7	1.21	0.14-4.36
Larynx	161-161.9	1	0.6	1.65	0.02-9.20
Respiratory tract	162-163.9	26	3.8	6.85	4.47-10.0
Lymphatic-haemopoietic	200-209.9	1	0.9	1.16	0.02-6.44
Other neoplasms		5	5.2	0.95	0.31-2.23
Benign neoplasms	210-239.9	1	0.4	2.57	0.03-14.3
Cardiovascular diseases	390-458.9	28	36.5	0.80	0.27-0.58
Respiratory tract diseases	460-519.9	102	7.5	13.63	11.1-16.5
Digestive tract diseases	520-577.9	18	5.4	3.33	1.97-5.26
Other diseases		36	5.3	6.81	4.77-9.43
Ill-defined conditions	780-796.9	-	2.1	-	-
Violent causes	800-999.9	-	3.1	-	-
All deaths		225	76.8	2.92	2.57-3.34

[a] O, observed deaths; E, expected deaths based on age and calendar-year-specific death rates of the male population of Italy (1961-81); SMR, standardized mortality ratio; 95% CI, 95% confidence interval; ICD code from the Ninth Revision of the *International Classification of Diseases*

SMRs computed on the basis of the mortality of the male population of Genoa (not shown) were similar to those estimated using the mortality rates for the entire Italian male population. Overall mortality (SMR=2.73, 95% CI=2.38-3.11), and mortality from all cancers (SMR=2.02, 95% CI=1.44-2.75), respiratory tract cancers (SMR=5.03, 95% CI=3.29-7.38), respiratory tract diseases (SMR=13.0, 95% CI=10.6-15.7), and other diseases (SMR=4.70, 95% CI=3.29-6.5), were all statistically significantly higher than the expected figures.

Mortality by age group is given in Table 5 and shows significant excess mortality from both malignant and non-malignant respiratory tract diseases, and from other diseases for the age groups 45-54, 55-64 and 65-74.

Indirect adjustment for smoking habits

Since smoking is a well known confounding factor, indirect adjustment (Axelson, 1978) was used in an attempt to estimate the expected mortality excess from respiratory tract cancers due to smoking habits. Table 6 shows the proportion of smokers, subjects who had never smoked and ex-smokers both in the group of silicotics and in the male population of Genoa (age group 25-75+). Given that smoking habits were unknown for 95 subjects (18.3%), two hypothetical distribu-

Table 5. Observed deaths from respiratory tract diseases, respiratory tract cancers, other diseases, and corresponding age-specific SMRs[a]

Causes of death	No. of deaths in age-group and SMR[b]					
	35-44	45-54	55-64	65-74	75+	Total
Respiratory cancers	-	5	10	11	-	26
	-	(24.87)	(7.63)	(5.89)	-	(6.85)
Respiratory diseases	-	11	25	45	21	102
	-	(70.07)	(19.64)	(14.38)	(7.20)	(13.63)
Other diseases	2	1	14	13	6	36
	(82.94)	(3.29)	(10.91)	(6.15)	(3.84)	(6.81)

[a] Computed by applying the mortality rates for the male population of Italy for the period 1 January 1961–31 December 1981
[b] In parentheses

Table 6. Smoking habits and crude estimates of respiratory tract cancer rate ratios attributable to smoking computed as a function of the proportion of smokers, ex-smokers, and non-smokers in the study group and in the reference population (i.e., the male population of Genoa)

Smoking habits	Study group				% of reference population
	%	No.	% based on assumption:[a]		
			(a)	(b)	
Smokers	47.3	246	53.39	65.56	48.8
Ex-smokers	20.4	106	26.47	20.40	15.1
Non-smokers[b]	14.0	73	20.12	14.03	36.1
Unknown	18.3	95	-	-	-
Rate ratio	0.971		1.119	1.302	1

[a] For explanation of assumptions (a) and (b), see text
[b] Subjects who had never smoked

tions of this subgroup were assumed: (a) all 95 subjects were assumed to be equally distributed between the three categories and (b) all 95 were assumed to belong to the smoking group. By setting the relative risk for non-smokers equal to 1, for ex-smokers equal to 4, and for smokers equal to 15, the excess mortality attributable solely to smoking was estimated to be 12% and 30% for assumptions (a) and (b) respectively.

Conclusions

The findings of this study clearly support both the expected association between silicosis and increased mortality from respiratory tract diseases as well as the suggested relationship between silicosis and mortality from cancers of the respiratory tract. Although the small size of the cohort precluded a meaningful stratified analysis by age at first exposure, duration of exposure, and length of follow-up, indirect adjustment for smoking habits (the major confounding factor in the development of respiratory cancers) has shown that smoking does not seriously affect the estimated SMR for lung cancer. Furthermore, since the study included only hospitalized subjects, cause-specific SMRs should be interpreted with caution. In particular, their extrapolation to the 'whole' population of workers exposed to mixed silica dusts should be avoided.

References

Archer, V.E., Gillam, J.D. & Wagner, J.K. (1986) Is silica or radon daughters the important factor in the excess lung cancer among underground miners? In: Goldsmith, D.F., Winn, D.M. & Shy, C.M., eds, *Silica, Silicosis and Cancer: Controversy in Occupational Medicine* (Cancer Research Monographs, Vol. 2), New York, NY, Praeger, pp. 375-384

Axelson, O. (1978) Aspects of confounding in occupational health epidemiology. *Scand. J. Work Environ. Health*, 4, 98-102

Finkelstein, M.M., Muller, J., Kusiak, R. & Suranyi, G. (1986) Follow-up of miners and silicotics in Ontario. In: Goldsmith, D.F., Winn, D.M. & Shy, C.M., eds, *Silica, Silicosis and Cancer: Controversy in Occupational Medicine* (Cancer Research Monographs, Vol. 2), New York, NY, Praeger, pp. 321-326

Goldsmith, D.F. & Guidotti, T.L. (1982) Does occupational exposure to silica cause lung cancer? *Am. J. Ind. Med.*, 3, 423-440

Goldsmith, D.F. & Guidotti, T.L. (1986) Combined silica exposure and cigarette smoking: a likely synergistic effect. In: Goldsmith, D.F., Winn, D.M. & Shy, C.M., eds, *Silica, Silicosis and Cancer: Controversy in Occupational Medicine* (Cancer Research Monographs, Vol. 2), New York, NY, Praeger, pp. 451-460

Heppleston, A.G. (1985) Silica, pneumoconiosis and carcinoma of the lung. *Am. J. Ind. Med.*, 7, 285-294

Hill, I.D. (1972) Computing man years at risk. *Br. J. Prev. Med.*, 26, 132-134

Holland, L.M., Wilson, J.S., Tillery, M.I. & Smith, D.M. (1986) Lung cancer in rats exposed to fibrogenic dusts. In: Goldsmith, D.F., Winn, D.M. & Shy, C.M., eds, *Silica, Silicosis and Cancer: Controversy in Occupational Medicine* (Cancer Research Monographs, Vol. 2), New York, NY, Praeger, pp. 267-279

IARC (1987) *IARC Monographs on the Evaluation of the Carcinogenic Risk of Chemicals to Humans*, Vol. 42, *Silica and Some Silicates*, Lyon, International Agency for Research on Cancer

Kurppa, K., Gudbergsson, H., Hannunkari, I., Koskinen, H., Hernberg, S., Koskela, R.-S. & Ahlmann, K. (1986) Lung cancer among silicotics in Finland. In: Goldsmith, D.F., Winn, D.M. & Shy, C.M., eds, *Silica, Silicosis and Cancer: Controversy in Occupational Medicine* (Cancer Research Monographs, Vol. 2), New York, NY, Praeger, pp. 311-320

Palmer, W.G. & Scott, W.D. (1981) Lung cancer in ferrous foundry workers: A review. *Am. Ind. Hyg. Assoc. J.*, 42, 329-340

Palmer, W.G. & Scott, W.D. (1986) Factors affecting lung cancer incidence in foundrymen. In: Goldsmith, D.F., Winn, D.M. & Shy, C.M., eds, *Silica, Silicosis and Cancer: Controversy in Occupational Medicine* (Cancer Research Monographs, Vol. 2), New York, NY, Praeger, pp. 45-56

Saffiotti, U. (1986) The pathology induced by silica in relation to fibrogenesis and carcinogenesis. In: Goldsmith, D.F., Winn, D.M. & Shy, C.M., eds, *Silica, Silicosis and Cancer: Controversy in Occupational Medicine* (Cancer Research Monographs, Vol. 2), New York, NY, Praeger, pp. 287-307

Santi, L. & Balestra, V. (1957) On the relationship between silicosis and lung cancer [in Italian]. *Accad. Med.*, II, 1-12

Sherson, D. & Iversen, E. (1986) Mortality among foundry workers in Denmark due to cancer and respiratory and cardiovascular diseases. In: Goldsmith, D.F., Winn, D.M. & Shy, C.M., eds, *Silica, Silicosis and Cancer: Controversy in Occupational Medicine* (Cancer Research Monographs, Vol. 2), New York, NY, Praeger, pp. 403-414

Thomas, T.L. (1982) A preliminary investigation of mortality among workers in the pottery industry. *Int. J. Epidemiol.*, *11*, 175-180

Wagner, M.M.F., Wagner, J.C., Davies, R. & Griffiths, D.M. (1980) Silica-induced malignant histiocytic lymphoma: incidence linked with strain of rat and type of silica. *Br. J. Cancer*, *41*, 908-917

Westerholm, P., Ahlmark, A., Massing, R. & Segelberg, I. (1986) Silicosis and lung cancer - a cohort study. In: Goldsmith, D.F., Winn, D.M. & Shy, C.M., eds, *Silica, Silicosis and Cancer: Controversy in Occupational Medicine* (Cancer Research Monographs, Vol. 2), New York, NY, Praeger, pp. 327-334

Zambon, P., Simonato, L., Mastrangelo, G., Winkelmann, R., Saia, B. & Crepet, M. (1986) A mortality study of workers compensated for silicosis during 1959 to 1963 in the Veneto Region of Italy. In: Goldsmith, D.F., Winn, D.M. & Shy, C.M., eds, *Silica, Silicosis and Cancer: Controversy in Occupational Medicine* (Cancer Research Monographs, Vol. 2), New York, NY, Praeger, pp. 367-374

Ziskind, M., Jones, R.N. & Weill, H. (1976) State of the art: Silicosis. *Am. Rev. Resp. Dis.*, *113*, 643-665

Lung cancer incidence among Swedish ceramic workers with silicosis

G. Tornling[1], C. Hogstedt[1,2] and P. Westerholm[3]

[1]*Department of Occupational Medicine,
Karolinska Hospital, Stockholm, Sweden*

[2]*National Institute of Occupational Health,
Stockholm, Sweden*

[3]*The Swedish Trade Union Confederation,
Stockholm, Sweden*

Summary. The incidence of lung cancer among 280 silicotic men working in the ceramics industry and notified to the Swedish Silicosis Registry has been investigated. During the study period 1958–83, the risk of lung cancer (nine cases) was double that expected based on national rates. There was no increased incidence of cancer at any other site. The results are in agreement with those of both animal and epidemiological studies of quartz exposure and point to an increased risk of lung cancer, especially among silicotics. Various possible explanations of this increased risk are discussed, but further studies are required.

Introduction

The causal relationship between occupational quartz exposure and silicosis is well established. In 1938, Anderson and Dible, in a discussion of the possibility that exposure to quartz could also increase the risk of lung tumours commented that: 'A group of cases of pulmonary carcinoma exists in which the organs contain an excess of silica and show histological evidence of silicotic fibrosis. The conclusion, we think, is that the role of the silicosis is etiological'.

More recently, two literature reviews covering both animal experiments and epidemiological investigations have been published. In the first of these, Goldsmith *et al.* (1982) concluded that an association appears to exist between exposure to quartz and lung cancer, whereas Heppleston (1985) came to the opposite conclusion. In 1984, a number of reports were presented at an international symposium on silica, silicosis and cancer that supported an association, but the main conclusion was that more studies were needed to confirm the hypothesis (Schneiderman & Winn, 1986). In 1986, IARC evaluated the carcinogenic risk of silica and some silicates, and found that the evidence that crystalline silica was a lung carcinogen for animals was conclusive but that the same was not true for humans (IARC, 1987).

A difficulty in the interpretation of several of the epidemiological investigations of the relationship between silica exposure and lung cancer has been that the individuals studied had often been exposed to other known lung carcinogens, such as polycyclic aromatic hydrocarbons (PAH) in foundries and ionizing radiation in mining (IARC, 1987). However, in the ceramics industry no such confounding exposures are known to occur, and the present investigation was therefore undertaken in order to study lung cancer incidence among ceramic workers with notified silicosis.

Exposure in the Swedish ceramics industry

Up to the 1950s, only sporadic measurements had been made of dust levels in Swedish workplaces with quartz exposure. However, the high frequency of silicosis in the early days of the ceramics industry is indicative of high quartz levels. In 1934 Bruce (1942) investigated the occurrence of silicosis in Swedish potteries and found prevalences ranging between 20 and 30%. For those employed for more than 20 years, the morbidity was over 50%. At a reinvestigation five years later, no new cases of silicosis were observed among 171 persons who had not been suffering from silicosis in 1934, indicating that the exposure situation had by that time improved.

The former Swedish National Institute for Public Health conducted an exposure investigation from 1947 to 1957 (Ahlmark & Öhman, 1957). Impinger measurements in a Swedish pottery showed mean levels of up to 20×10^6 particles per cubic foot (560 per m^3) in certain processes but, for the majority, measurements made in the breathing zone gave a figure of less than 10×10^6 particles per cubic foot.

Between 1963 and 1967, information on current silicosis hazards in Swedish industry was systematically collected (the Silicosis Investigation) (Ahlmark, 1967). In the ceramics industry, the mean dust exposure, as determined by stationary measurements, was 4.9 mg/m^3, 17% of the dust being crystalline silica and 30% particles less than 5 μm in size.

The Silicosis Investigation was followed by the Silicosis Project, which lasted from 1968 to 1971 and was conducted by the National Institute of Occupational Medicine (Gerhardsson *et al.*, 1974). This showed that talc had been used in four factories producing sanitary ware, but no exposure measurements were made, and it is not clear whether the talc was fibrous or not. In potteries, the mean dust exposure in the breathing zone was 7.1 mg/m^3 (median 2.8 mg/m^3), the quartz content was 9% and 39% of particles were less than 5 μm in size. In other ceramic industries, the mean dust exposure in the breathing zone was 8.2 mg/m^3 (median 2.9 mg/m^3), the quartz content was 11% and 16% of particles were less than 5 μm in size.

Study population

The Swedish Pneumoconiosis Registry was established in 1953, and has collected cases notified to the National Social Insurance Board and to private insurance companies from 1931 onwards. At the Registry, the information from each case has been recorded systematically. All radiographic examinations and medical records have been reviewed by specialists. A total of 314 males and 36 females from the ceramics industry have been notified and accepted as silicotics for compensation purposes. Because of their small number, females were excluded from further analysis in this study.

The study population for the cancer incidence calculations consists of 280 male ceramic workers from 19 different factories who had contracted silicosis and were alive on 1 January 1958 when the Swedish Cancer Registry was started. Altogether, they represent 4247 person-years during the study period 1958–83. The members of the cohort were born during the period 1875–1934, and half of them were born before 1910. They were generally first employed in the ceramics industry before the age of 25, and silicosis was seldom detected until 30 years later.

For the mortality calculations, the study population was extended to include those male ceramic workers who were alive on 1 January 1951. For the study period 1951–85, they represent 5695 person-years.

Methods

The vital status of the members of the cohort was established by linking the ten-digit identification number based on date of birth for each individual with the census register of all living persons in Sweden at the time of the follow-up, the death register of all deceased persons in Sweden, and the emigration register. By this means all but one of the subjects could be identified as alive or deceased during the study period.

The cancer morbidity for 1958–83 was established by linking the identification number of all individuals with the National Cancer Register; this was started in 1958 and more than 95% of all diagnosed malignancies in Sweden are notified to it (Mattsson, 1984).

The expected number of malignancies was calculated by multiplying person-years of observation within five-year age categories during each year of the study period by site-specific and sex-specific national rates. The cancer incidence, the relative risk (SMR, standardized mortality ratio) and the 95% confidence interval (CI), based on a Poisson distribution, were calculated by means of a computer programme developed at the University of Linköping (EPILIN programme package). Calculations were also carried out for latency periods of ten and 20 years after the diagnosis of silicosis, since it has been suggested that fibrosis may lead to the development of cancer.

Causes of death were taken from the death certificates, and calculations similar to those for cancer incidence were performed for cause of death during the study period 1951–85.

Results

The analysis showed that the overall mortality in the study population was increased (218 observed *versus* 158.2 expected; SMR 138; 95% CI 120–157) due to excess mortality from respiratory tuberculosis (18 observed *versus* 0.9 expected; SMR 1932; 95% CI 1144–3054) and non-malignant respiratory diseases (67 observed *versus* 9.0 expected; SMR 746; 95% CI 577–947).

There was no increased overall incidence of malignant diseases in the study population (Table 1). However, nine cases of lung cancer were observed versus 4.8 expected, the SMR being 188 (95% CI 85–356). With a latency period of ten years from the discovery of silicosis, the SMR was 236 (95% CI 107–448), and with a latency period of 20 years there was a further increase in the SMR to 267 (95% CI 98–582). There was no excess of cancer at any other site either overall or after latency periods of ten and 20 years.

Table 1. Number of malignant neoplasms in the study population and SMR with 95% confidence interval[a]

Site	All cohort				>10 years after detection of silicosis			
	O	E	SMR	95% CI	O	E	SMR	95% CI
All sites	41	43.8	94	67–126	36	35.9	100	70–138
Stomach	4	4.3	92	25–236	4	3.6	112	30–288
Colon	2	3.3	61	7–221	2	2.7	74	8–267
Pancreas	2	1.8	113	13–408	1	1.5	69	1–385
Lung	9	4.8	188	85–356	9	3.8	236	107–448
Bladder and urinary system (excluding kidney)	3	2.8	109	22–317	3	2.3	133	27–389
Blood	5	3.3	152	49–355	5	2.6	190	61–444

[a] O, observed; E, expected; SMR, standardized mortality ratio; 95% CI, 95% confidence interval

Dates of birth, periods of employment in the ceramics industry, dates of detection of silicosis and cancer diagnoses for the nine lung cancer cases are presented in Table 2. There were three squamous-cell carcinomas, two small-cell carcinomas, two adenocarcinomas, one alveolar-cell carcinoma and one undifferentiated carcinoma. Lung cancer was diagnosed 36–72 years after initial exposure to quartz and 11–32 years after the detection of silicosis. The individuals with lung cancer had worked at six different factories, and there was no concentration of lung cancer cases in factories making a particular product.

Discussion

The results of this study show that the incidence of lung cancer among silicotic workers in the Swedish ceramics industry is twice as great as expected. The diagnosis of silicosis should be extremely reliable as all radiographic examinations and medical records are routinely reviewed by qualified and experienced specialists at the Pneumoconiosis Registry.

Smoking habits should obviously be taken into account in any study of lung cancer, but we have no relevant information on the population in the present study. If the proportion of smokers in the study population were to be different from that in the general population, smoking would be a true confounding factor.

Table 2. Date of birth, period of employment in the ceramics industry, dates of detection of silicosis and cancer diagnosis and histological type for the nine lung cancer cases

Date of birth	Period of employment	Date of detection of silicosis	Date of cancer diagnosis	Histological type
1890	1904-57	1956	1976	Undifferentiated
1893	1906-61	1934	1966	Squamous-cell
1900	1923-37	1946	1966	Small-cell
1904	1918-37	1934	1965	Alveolar-cell
1907	1942-57	1956	1978	Squamous-cell
1907	1940-73	1971	1982	Squamous-cell
1908	1925-49	1956	1973	Small-cell
1911	1930-45	1963	1980	Adenocarcinoma
1918	1934-50	1955	1982	Adenocarcinoma

This might be the case, for example, either if smoking increased the risk of developing silicosis, or if silicotics with smoking-related respiratory morbidity are more likely to be included in the Pneumoconiosis Registry. However, an association between smoking and silicosis has not been demonstrated by any study. In addition, all employees in the ceramics industry have undergone regular chest X-rays, whether respiratory symptoms were present or not, and the Pneumoconiosis Registry has been notified when silicosis has been found, even if the individual concerned has not applied for compensation.

It has been suggested that fibrous talc may be carcinogenic. However, talc was used in only four of the 19 factories, and the lung cancer cases were not concentrated in these factories.

The cohort is small, and the results should be evaluated in relation to those of other studies, primarily on silicotics working in the ceramics industry. Data from the Nordic census-based occupational mortality and cancer incidence registers show no consistent pattern for lung cancer within the glass, porcelain, ceramics and tile products industries (Lynge et al., 1986). For all the groups combined, a significant excess was seen in Norway and a slight excess in Finland, whereas the overall risk in both Sweden and Denmark was close to the national average. No significant excess of lung cancer was seen in the ceramics industry. However, this lack of association between silica and lung cancer in record-linkage studies based on census data should be interpreted with great caution, since the working classifications are broad and the information on occupation refers only to one particular time, i.e., the true risk is underestimated because of misclassification and dilution bias.

Thomas (1982) and Thomas et al. (1986) reported an increased mortality from lung cancer among pottery workers (O/E=1.21, $p<0.01$), but the increase was restricte to those who had worked in factories making sanitary ware (O/E=1.80, $p<0.01$) where talc had been used for the slip casting of large ceramic pieces. In a cohort mortality study on three plants producing ceramic plumbing fixtures, Thomas and Stewart (1987) subdivided the cohort according to silica and talc exposure. Those who had been exposed to high levels of silica dust but without exposure to talc had a non-significant SMR of 1.37. Those who had been exposed to non-fibrous talc in addition to high levels of silica had a significant 2.54-fold excess of lung cancer.

An Italian case–referent study on lung cancer among workers in the ceramics industry where allowance was made for age, period of death, and smoking (Forastiere et al., 1986), showed a higher lung cancer risk for workers in the ceramics industry than for those in occupations free from silica exposure (Mantel–Haenszel rate ratio 2.0; 95% CI 1.1–3.5). The increased risk was mainly due to a rate ratio of 3.9 (95% CI 1.8–8.3) for silicotics; it was only 1.4 (95% CI 0.7–2.8) for non-silicotics.

Animal experiments have shown a significant increase in the frequency of lung tumours in rats after exposure to quartz by intratracheal instillation (Groth et al., 1986) as well as inhalation (Johnson et al., 1987).

Overall, the studies suggest an increased incidence of lung cancer due to quartz exposure, especially among silicotics, even if the possible confounding effect of talc is taken into account.

The mechanism whereby quartz could exert a carcinogenic effect is not clear. Since it has a cytotoxic effect *in vitro* (Langer & Nolan, 1986), it may be directly carcinogenic. Silicotics seem to be at greater risk of lung cancer than non-silicotics when exposed to quartz, possibly because their exposure to quartz has been greater than that of the latter. However, a causal relationship between the fibrotic lesions in the lung and lung cancer is also possible.

A high incidence of bronchial carcinoma among individuals with other fibrotic lung diseases unrelated to dust exposure, such as cryptogenic fibrosing alveolitis and pulmonary systemic sclerosis (Turner-Warwick, 1978) has been demonstrated. Fibrotic lesions, however they originate, might impair the pulmonary clearance mechanism for various toxic and carcinogenic substances. A synergistic rather than additive effect between quartz exposure and smoking in lung cancer etiology has been proposed by Goldsmith and Guidotti (1986), but the small number of studies so far carried out make it difficult to predict the magnitude of any such effect.

In conclusion, the results obtained from this cohort of silicotic ceramic workers support the findings of an increased risk of lung cancer. Whether this increased risk is caused directly by quartz dust, by the impairment of important lung functions or by some other fibrosis-related mechanism remains unclear. Further experimental and epidemiological studies on this subject are therefore necessary.

Acknowledgements

We are indebted to Annika Gustavsson at the National Institute of Occupational Health for data collection and computer assistance. This study was supported by grant 86-1391 from the Swedish Work Environment Fund.

References

Ahlmark, A. (1967) *Silicosis in Sweden* [in Swedish], Stockholm, Arbetarskyddsstyrelsen (Studia Laboris et Salutis)

Ahlmark, A. & Öhman, H. (1957) *Silicosis and its control* [in Swedish]. Solna, Föreningen för Arbetarskydd

Anderson, C.S. & Dible, J.H. (1938) Silicosis and carcinoma of lung. *J. Hyg.*, *38*, 185-204

Bruce, T. (1942) Silicosis as an occupational disease in Sweden [in German]. *Acta Med. Scand.*, *110*, Suppl. 129

Forastiere, F., Lagorio, S., Michelozzi, P., Cavariani, F., Borgia, P., Arcà, M., Perucci, C. & Axelson, O. (1986) Silica, silicosis and lung cancer among ceramic workers: a case-referent study. *Am. J. Ind. Med.*, *10*, 363-370

Gerhardsson, G., Engman, L., Andersson, A., Isaksson, G., Magnusson, E. & Sundquist, S. (1974) *Silicosis Project: Final report* [in Swedish], Stockholm, Arbetarskyddsstyrelsen (Undersökningsrapport AMT 103/74-2)

Goldsmith, D.F. & Guidotti, T.L. (1986) Combined silica exposure and cigarette smoking: A likely synergistic effect. In: Goldsmith, D.F., Winn, D.M. & Shy, C.M., eds, *Silica, Silicosis and Cancer: Controversy in Occupational Medicine* (Cancer Research Monographs, Vol. 2), New York, NY, Praeger, pp. 451-459

Goldsmith, D.F., Guidotti, T.L. & Johnston, D.R. (1982) Does occupational exposure to silica cause lung cancer? *Am. J. Ind. Med.*, *3*, 423-440

Groth, D.H., Stettler, L.E., Platek, S.F., Lal, J.B. & Burg, J.R. (1986) Lung tumours in rats treated with quartz by intratracheal installation. In: Goldsmith, D.F., Winn, D.M. & Shy, C.M., eds, *Silica, Silicosis and Cancer: Controversy in Occupational Medicine* (Cancer Research Monographs, Vol. 2), New York, NY, Praeger, pp. 243-253

Heppleston, A.G. (1985) Silica, pneumoconiosis and carcinoma of the lung. *Am. J. Ind. Med.*, *7*, 285-294

IARC (1987) *IARC Monographs on the Evaluation of the Carcinogenic Risk of Chemicals to Humans:* Vol. 42, *Silica and Some Silicates*, Lyon, International Agency for Research on Cancer

Johnson, N.F., Smith, D.M., Sebring, R. & Holland, L.M. (1987) Silica-induced alveolar cell tumors in rats. *Am. J. Ind. Med.*, *11*, 93-107

Langer, A.M. & Nolan, R.P. (1986) Physiochemical properties of quartz controlling biological activity. In: Goldsmith, D.F., Winn, D.M. & Shy, C.M., eds, *Silica, Silicosis and Cancer: Controversy in Occupational Medicine* (Cancer Research Monographs, Vol. 2), New York, NY, Praeger, pp. 125-135

Lynge, E., Kurppa, K., Kristofersen, L., Malker, H. & Sauli, H. (1986) Silica dust and lung cancer: results from the Nordic occupational mortality and cancer incidence registers. *J. Natl Cancer Inst.*, *77*, 883-889

Mattsson, B. (1984) *Cancer Registration in Sweden*, Thesis, Stockholm

Schneiderman, M. & Winn, D.M. (1986) Where are we with the SiO_2 and cancer issue? In: Goldsmith, D.F., Winn, D.M. & Shy, C.M., eds, *Silica, Silicosis and Cancer: Controversy in Occupational Medicine* (Cancer Research Monographs, Vol. 2), New York, NY, Praeger, pp. 523-531

Thomas, T.L. (1982) A preliminary investigation of mortality among workers in the pottery industry. *Int. J. Epidemiol.*, *11*, 175-180

Thomas, T.L. & Stewart, P.A. (1987) Mortality from lung cancer and respiratory disease among pottery workers exposed to silica and talc. *Am. J. Epidemiol.*, *125*, 35-43

Thomas, T.L., Stewart, P.A. & Blair, A. (1986) Nonfibrous dust and cancer: Studies at the National Cancer Institute. In: Goldsmith, D.F., Winn, D.M. & Shy, C.M., eds, *Silica, Silicosis and Cancer: Controversy in Occupational Medicine* (Cancer Research Monographs, Vol. 2), New York, NY, Praeger, pp. 441-450

Turner-Warwick, M. (1978) *Immunology of the Lung*, London, Edward Arnold

Index

Anthrasilicosis in pneumoconiosis patients in Japan 96
Asbestos 35

Benzo[a]pyrene 69, 70
Bronchial carcinoma and fibrotic lung disease 118

Carcinogenesis, multistage model 70
Carcinogenicity of silica 1, 4, 29-30, 43, 83, 113
    mechanisms 50-51, 118
Case-control study, lung cancer cases in Japan 97
Case-referent study in Civitacastellana 22-27
Case-referent study in Montreal
    case groups and control groups 32-34
    design 30
    ethnic group as confounder and effect modifier 31-32, 35
    excess risk for lung cancer 36
    exposure to silica 34
    face-to-face interviews 30
    French Canadians 31-32
    methods of analysis 34-35
    odds ratios for smoking-silica combinations 36
    questionnaire 31
    retrospective exposure assessment 31, 39
    smoking as confounder 35
    socioeconomic status as confounder 35
    stomach cancer 36
Ceramic plumbing-fixture industry
    see Ceramics industry
Ceramic plumbing fixtures
    see Sanitary ware, manufacture in the United States
Ceramics industry
    case-referent study on lung cancer 118
    respiratory hazards 23
    see also Ceramic workers; Pottery workers; Sanitary-ware workers
Chromate pigments 23, 24, 27
Civitacastellana
    case-referent study 22-27
    lung cancer cases 22
Construction industry 36, 70

Denmark, lung cancer cases 9
Diesel exhaust fumes 23, 24, 27
Dust exposure
    duration, and lung cancer 99
    and lung cancer mortality 93
    mean, in Swedish potteries 114
    silica, in Swedish miners 13
    see also Quartz dust, exposure of granite workers in Finland; Silica exposure
Dusts
    inert 69, 71
    inorganic, association with non-adenocarcinoma of the lung 40

Excavation work 15
    lung cancer risk 16

Fibrotic lung disease and bronchial carcinoma 118
Finland
    follow-up study of granite workers
        cohort 44
        deaths due to lung cancer 47
        exposure to quartz dust 45
        mortality of cohort 45
        potential confounders 45
        smoking habits 46, 49
        statistical analysis 46
    lung cancer deaths 9
Foundry work 10
Foundry workers
    lung cancer 10
    stomach cancer 69
French Canadians, in Montreal case-referent study 31-32

German Democratic Republic
    cancer in workers in slate quarries 57, 58
    silicosis in workers in slate quarries 63
Glass, porcelain, ceramics and tile industry, lung cancer risk 13
Glass workers, excess risk 13
Granite workers 15
    follow-up study in Finland
        cohort 44
        deaths due to lung cancer 47
        exposure to quartz dust 45
        mortality of cohort 45
        potential confounders 45
        smoking habits 46, 49
        statistical analysis 46

Healthy worker effect 68, 71, 90, 94

Indices of relative mortality 10
Inert dusts 69, 71
Inorganic dusts, association with non-adenocarcinoma of the lung 40
Ionizing radiation 114
see also Radon daughters

Japan
  case-control study for lung cancer cases 97
  pneumoconiosis patients
    with anthrasilicosis 96
    elevated lung cancer mortality 102
    lung cancer risk 95
    with silicosis 96
    smoking and lung and stomach cancer 98
    study population 96, 99
  see also 'Rosai' hospital system

Lung cancer 3-4
  association with long-term high-level silica exposure 40
  association with silicosis 62, 106
  attributable percentage risk due to silica exposure 38
  case-control study in Japan 97
  case-referent study among workers in the ceramics industry 118
  cases in Civitacastellana 22
  and cigarette smoking 85
  deaths in follow-up study of granite workers in Finland 47
  deaths in pottery workers 92-93, 117
  in Denmark 9
  and dust exposure 93, 99
  elevated mortality among pneumoconiosis patients in Japan 102
  elevated odds ratios in Montreal study 36
  etiological role of silicosis 113
  excess mortality in pottery workers 90, 94
  excess mortality in sanitary-ware workers 76, 77
  excess risk among silicotics 63
  in Finland 9
  among foundry workers 10, 69
  frequency by radiographic category of pneumoconiosis 99
  frequency among silicotics 99
  higher risk of silicosis patients 103
  histological types 99
  among miners 10, 13
  in Norway 8
  in occupational groups with potential exposure to silica 7
  relationship with cigarette smoking 85
  risk in glass, porcelain, ceramics and tile industry 13
  risk in miners 10, 13
  risk in pottery workers 13, 80
  risk in sanitary-ware workers 79
  silicosis patients, higher risk 103
  in silicotic ceramic workers 110
  and smoking in pneumoconiosis patients in Japan 98, 102
  in Sweden 9
  in workers in slate quarries in the German Democratic Republic 57, 58
  see also Lung tumours after exposure to quartz by intratracheal instillation; Non-adenocarcinoma of the lung, association with inorganic dusts; Respiratory tract cancers
Lung tumours after exposure to quartz by intratracheal instillation 118
  see also Lung cancer; Non-adenocarcinoma of the lung, association with inorganic dusts; Respiratory tract cancers

Miners, lung cancer risk 10, 13
Montreal, population-based case-referent study
  case groups and control groups 32-34
  design 30
  ethnic group as confounder and effect modifier 31-32, 35
  excess risk for lung cancer
  exposure to silica 34
  face-to-face interviews 30
  French Canadians 31-32
  methods of analysis 34-35
  odds ratios for smoking-silica combinations 36
  questionnaire 31
  retrospective exposure assessment 31, 39
  exposure as confounder 35
  socioeconomic status as confounder 35
  stomach cancer 36

Non-adenocarcinoma of the lung, association with inorganic dusts 40
Non-malignant respiratory disease
  excess mortality in pottery workers 76, 77
  mortality in sanitary-ware workers 77, 78
Non-silicotics 22, 23, 24

Nordic occupational mortality and cancer
    incidence registers 7-8
Norway, lung cancer deaths 8

Odds ratios for smoking-silica
    combinations in Montreal
    population-based case-referent
    study 36

PAH *see* Polycyclic aromatic
    hydrocarbons
Pneumoconiosis, radiographic category
    and frequency of lung cancer 99
    and lung cancer risk 103
Pneumoconiosis patients in Japan
    with anthrasilicosis 96
    elevated lung cancer mortality 102
    lung cancer risk 95
    with silicosis 96
    smoking and lung and stomach cancer
        98
    study population 96, 99
Pneumoconiosis silicotica 15
Polycyclic aromatic hydrocarbons 69, 70,
    114
Pottery workers
    excess lung cancer mortality 90
    excess mortality 76, 89
    excess rectal cancer mortality 90
    excess stomach cancer mortality 90
    follow-up analysis 89-93
    lung cancer deaths 92-93
    main follow-up study 87-89
    methods used in survey in Britain
        84-85
    pilot follow-up study 86-87
    proportional mortality ratio study 75
    smoking 91, 92
    studies 22
    study sample in Britain 85
    survey of respiratory disease in Britain
        83
    see also Ceramics industry; Ceramic
        workers; Sanitary-ware workers;
        Silicotic ceramic workers

Quarrymen 22, 23
Quartz concentrations in the workplace
    93
Quartz dust, exposure of granite workers
    in Finland 45

Radon daughters 13
    see also Ionizing radiation
Rectal cancer
    excess mortality in pottery workers
        90, 94
    in workers in slate quarries in the
        German Democratic Republic 57

Relative mortality, indices 10
Respiratory tract cancers 106
    mortality and silicosis 109
    in silicotics 107
    see also Lung cancer; Lung tumours
        after exposure to quartz by
        intratracheal instillation;
        Non-adenocarcinoma of the lung,
        association with inorganic dusts
Respiratory tract diseases
    association between silicosis and
        increased mortality 109
    excess mortality in silicotics 107
Retrospective exposure assessment, in
    Montreal population-based
    case-referent study 31, 39
'Rosai' hospital system 95

Sand and gravel pits 15
Sanitary ware, manufacture in the United
    States 77
Sanitary-ware workers in the United
    States
    lung cancer mortality 76, 77, 78
    lung cancer risk 79
    mortality from non-malignant
        respiratory disease 77, 78
    silica exposure 78-79
    see also Ceramics industry; Ceramic
        workers; Pottery workers; Silicotic
        ceramic workers
Selection biases, in Vienna prospective
    study 71
Silica, carcinogenicity, 1, 4, 29-30, 83
    mechanisms 50-51
Silica exposure
    association with lung cancer 40
    attributable percentage risk of lung
        cancer 38
    in Montreal population-based
        case-referent study 34
    see also Dust exposure; Quartz dust,
        exposure of granite workers in
        Finland
Silicosis 50
    association with increased mortality
        from respiratory tract diseases 109
    association with lung cancer 62, 106
    etiological role in lung cancer 113
    and lung cancer risk 103
    and mortality from cancers of the
        respiratory tract 109
    in pneumoconiosis patients in Japan
        96
    among slate quarry workers in the
        German Democratic Republic 63
    in Swedish potteries 114
    see also following entries
Silicosis Investigation 114

Silicosis Project 114
Silicotic ceramic workers
    lung cancer 27, 116
    overall mortality 115
    smoking habits 116-117
    study population 115
    see also Ceramics industry; Ceramic workers; Pottery workers; Sanitary-ware workers; Silicosis; Silicotics
Silicotics 22, 23, 24, 26, 43
    epidemiological studies 55
    excess mortality 107
    excess risk for lung cancer 63
    historical cohort mortality study 106-108
    respiratory tract cancers 107
    see also Silicosis; Silicotic ceramic workers
Silicotuberculosis 15, 107
Silicotuberculotics, Austrian 61, 66
Slate quarries in the German Democratic Republic 56
    mortality among workers 57, 63
    risk for lung cancer among workers 63
    study population 56-57
Smoking 23, 24, 26, 27, 40
    habits of cohort in follow-up study of granite workers in Finland 46, 49
    and increase in lung cancer deaths in pneumoconiosis patients in Japan 102
    indirect adjustment 108-109
    and lung and stomach cancer in pneumoconiosis patients in Japan 98
    odds ratios for combinations with silica in Montreal study 36
    in pottery workers 91, 92, 94
    relationship with lung cancer, adjustment to number of expected deaths 85, 92
    in silicotic ceramic workers 116-117
Stomach cancer
    elevated rates in dusty trades 71-72
    excess mortality in pottery workers 90, 94
    in iron foundry workers 69
    in Montreal study 36, 38, 40
    and smoking in pneumoconiosis patients in Japan 98
    in workers in slate quarries in the German Democratic Republic 57
Stone cutters 15
Stone quarries 15
Sweden, lung cancer cases 9
Swedish Pneumoconiosis Registry 114

Talc 23, 24, 26, 114
    fibrous 77, 117
    non-fibrous 77, 78, 70, 81

United States, manufacture of sanitary ware 77

Vienna prospective study
    cancer deaths and standardized mortality ratios 69
    confounding exposures 70-71
    excess mortality in workers exposed to dust 68
    methods 67
    selection biases 71
    study population 66
Vienna, screening examinations for workers 65

Workers compensated for silicosis 2
    excess lung cancer mortality 63
    mortality 57
Workers exposed to silica and to carcinogens 2-3
Workers exposed to silica but not to carcinogens 3-4

PUBLICATIONS OF THE INTERNATIONAL AGENCY FOR RESEARCH ON CANCER
Scientific Publications Series

(Available from Oxford University Press through local bookshops)

No. 1 Liver Cancer
1971; 176 pages (*out of print*)

No. 2 Oncogenesis and Herpesviruses
Edited by P.M. Biggs, G. de-Thé and L.N. Payne
1972; 515 pages (*out of print*)

No. 3 N-Nitroso Compounds: Analysis and Formation
Edited by P. Bogovski, R. Preussman and E.A. Walker
1972; 140 pages (*out of print*)

No. 4 Transplacental Carcinogenesis
Edited by L. Tomatis and U. Mohr
1973; 181 pages (*out of print*)

No. 5/6 Pathology of Tumours in Laboratory Animals, Volume 1, Tumours of the Rat
Edited by V.S. Turusov
1973/1976; 533 pages; £50.00

No. 7 Host Environment Interactions in the Etiology of Cancer in Man
Edited by R. Doll and I. Vodopija
1973; 464 pages; £32.50

No. 8 Biological Effects of Asbestos
Edited by P. Bogovski, J.C. Gilson, V. Timbrell and J.C. Wagner
1973; 346 pages (*out of print*)

No. 9 N-Nitroso Compounds in the Environment
Edited by P. Bogovski and E.A. Walker
1974; 243 pages; £21.00

No. 10 Chemical Carcinogenesis Essays
Edited by R. Montesano and L. Tomatis
1974; 230 pages (*out of print*)

No. 11 Oncogenesis and Herpesviruses II
Edited by G. de-Thé, M.A. Epstein and H. zur Hausen
1975; Part I: 511 pages
Part II: 403 pages; £65.00

No. 12 Screening Tests in Chemical Carcinogenesis
Edited by R. Montesano, H. Bartsch and L. Tomatis
1976; 666 pages; £45.00

No. 13 Environmental Pollution and Carcinogenic Risks
Edited by C. Rosenfeld and W. Davis
1975; 441 pages (*out of print*)

No. 14 Environmental N-Nitroso Compounds. Analysis and Formation
Edited by E.A. Walker, P. Bogovski and L. Griciute
1976; 512 pages; £37.50

No. 15 Cancer Incidence in Five Continents, Volume III
Edited by J.A.H. Waterhouse, C. Muir, P. Correa and J. Powell
1976; 584 pages; (*out of print*)

No. 16 Air Pollution and Cancer in Man
Edited by U. Mohr, D. Schmähl and L. Tomatis
1977; 328 pages (*out of print*)

No. 17 Directory of On-going Research in Cancer Epidemiology 1977
Edited by C.S. Muir and G. Wagner
1977; 599 pages (*out of print*)

No. 18 Environmental Carcinogens. Selected Methods of Analysis. Volume 1: Analysis of Volatile Nitrosamines in Food
Editor-in-Chief: H. Egan
1978; 212 pages (*out of print*)

No. 19 Environmental Aspects of N-Nitroso Compounds
Edited by E.A. Walker, M. Castegnaro, L. Griciute and R.E. Lyle
1978; 561 pages (*out of print*)

No. 20 Nasopharyngeal Carcinoma: Etiology and Control
Edited by G. de-Thé and Y. Ito
1978; 606 pages (*out of print*)

No. 21 Cancer Registration and its Techniques
Edited by R. MacLennan, C. Muir, R. Steinitz and A. Winkler
1978; 235 pages; £35.00

No. 22 Environmental Carcinogens. Selected Methods of Analysis. Volume 2: Methods for the Measurement of Vinyl Chloride in Poly(vinyl chloride), Air, Water and Foodstuffs
Editor-in-Chief: H. Egan
1978; 142 pages (*out of print*)

No. 23 Pathology of Tumours in Laboratory Animals. Volume II: Tumours of the Mouse
Editor-in-Chief: V.S. Turusov
1979; 669 pages (*out of print*)

Prices, valid for January 1990, are subject to change without notice

List of IARC Publications

No. 24 Oncogenesis and Herpesviruses III
Edited by G. de-Thé, W. Henle and F. Rapp
1978; Part I: 580 pages, Part II: 512 pages (*out of print*)

No. 25 Carcinogenic Risk. Strategies for Intervention
Edited by W. Davis and C. Rosenfeld
1979; 280 pages (*out of print*)

No. 26 Directory of On-going Research in Cancer Epidemiology 1978
Edited by C.S. Muir and G. Wagner
1978; 550 pages (*out of print*)

No. 27 Molecular and Cellular Aspects of Carcinogen Screening Tests
Edited by R. Montesano, H. Bartsch and L. Tomatis
1980; 372 pages; £29.00

No. 28 Directory of On-going Research in Cancer Epidemiology 1979
Edited by C.S. Muir and G. Wagner
1979; 672 pages (*out of print*)

No. 29 Environmental Carcinogens. Selected Methods of Analysis. Volume 3: Analysis of Polycyclic Aromatic Hydrocarbons in Environmental Samples
Editor-in-Chief: H. Egan
1979; 240 pages (*out of print*)

No. 30 Biological Effects of Mineral Fibres
Editor-in-Chief: J.C. Wagner
1980; **Volume 1:** 494 pages; **Volume 2:** 513 pages; £65.00

No. 31 N-Nitroso Compounds: Analysis, Formation and Occurrence
Edited by E.A. Walker, L. Griciute, M. Castegnaro and M. Börzsönyi
1980; 835 pages (*out of print*)

No. 32 Statistical Methods in Cancer Research. Volume 1. The Analysis of Case-control Studies
By N.E. Breslow and N.E. Day
1980; 338 pages; £20.00

No. 33 Handling Chemical Carcinogens in the Laboratory
Edited by R. Montesano, *et al.*
1979; 32 pages (*out of print*)

No. 34 Pathology of Tumours in Laboratory Animals. Volume III. Tumours of the Hamster
Editor-in-Chief: V.S. Turusov
1982; 461 pages; £39.00

No. 35 Directory of On-going Research in Cancer Epidemiology 1980
Edited by C.S. Muir and G. Wagner
1980; 660 pages (*out of print*)

No. 36 Cancer Mortality by Occupation and Social Class 1851–1971
Edited by W.P.D. Logan
1982; 253 pages; £22.50

No. 37 Laboratory Decontamination and Destruction of Aflatoxins B_1, B_2, G_1, G_2 in Laboratory Wastes
Edited by M. Castegnaro, *et al.*
1980; 56 pages; £6.50

No. 38 Directory of On-going Research in Cancer Epidemiology 1981
Edited by C.S. Muir and G. Wagner
1981; 696 pages (*out of print*)

No. 39 Host Factors in Human Carcinogenesis
Edited by H. Bartsch and B. Armstrong
1982; 583 pages; £46.00

No. 40 Environmental Carcinogens. Selected Methods of Analysis. Volume 4: Some Aromatic Amines and Azo Dyes in the General and Industrial Environment
Edited by L. Fishbein, M. Castegnaro, I.K. O'Neill and H. Bartsch
1981; 347 pages; £29.00

No. 41 N-Nitroso Compounds: Occurrence and Biological Effects
Edited by H. Bartsch, I.K. O'Neill, M. Castegnaro and M. Okada
1982; 755 pages; £48.00

No. 42 Cancer Incidence in Five Continents, Volume IV
Edited by J. Waterhouse, C. Muir, K. Shanmugaratnam and J. Powell
1982; 811 pages (*out of print*)

No. 43 Laboratory Decontamination and Destruction of Carcinogens in Laboratory Wastes: Some N-Nitrosamines
Edited by M. Castegnaro, *et al.*
1982; 73 pages; £7.50

No. 44 Environmental Carcinogens. Selected Methods of Analysis. Volume 5: Some Mycotoxins
Edited by L. Stoloff, M. Castegnaro, P. Scott, I.K. O'Neill and H. Bartsch
1983; 455 pages; £29.00

No. 45 Environmental Carcinogens. Selected Methods of Analysis. Volume 6: N-Nitroso Compounds
Edited by R. Preussmann, I.K. O'Neill, G. Eisenbrand, B. Spiegelhalder and H. Bartsch
1983; 508 pages; £29.00

No. 46 Directory of On-going Research in Cancer Epidemiology 1982
Edited by C.S. Muir and G. Wagner
1982; 722 pages (*out of print*)

List of IARC Publications

No. 47 Cancer Incidence in Singapore 1968–1977
Edited by K. Shanmugaratnam, H.P. Lee and N.E. Day
1983; 171 pages (*out of print*)

No. 48 Cancer Incidence in the USSR (2nd Revised Edition)
Edited by N.P. Napalkov, G.F. Tserkovny, V.M. Merabishvili, D.M. Parkin, M. Smans and C.S. Muir
1983; 75 pages; £12.00

No. 49 Laboratory Decontamination and Destruction of Carcinogens in Laboratory Wastes: Some Polycyclic Aromatic Hydrocarbons
Edited by M. Castegnaro, *et al.*
1983; 87 pages; £9.00

No. 50 Directory of On-going Research in Cancer Epidemiology 1983
Edited by C.S. Muir and G. Wagner
1983; 731 pages (*out of print*)

No. 51 Modulators of Experimental Carcinogenesis
Edited by V. Turusov and R. Montesano
1983; 307 pages; £22.50

No. 52 Second Cancers in Relation to Radiation Treatment for Cervical Cancer: Results of a Cancer Registry Collaboration
Edited by N.E. Day and J.C. Boice, Jr
1984; 207 pages; £20.00

No. 53 Nickel in the Human Environment
Editor-in-Chief: F.W. Sunderman, Jr
1984; 529 pages; £41.00

No. 54 Laboratory Decontamination and Destruction of Carcinogens in Laboratory Wastes: Some Hydrazines
Edited by M. Castegnaro, *et al.*
1983; 87 pages; £9.00

No. 55 Laboratory Decontamination and Destruction of Carcinogens in Laboratory Wastes: Some N-Nitrosamines
Edited by M. Castegnaro, *et al.*
1984; 66 pages; £7.50

No. 56 Models, Mechanisms and Etiology of Tumour Promotion
Edited by M. Börzsönyi, N.E. Day, K. Lapis and H. Yamasaki
1984; 532 pages; £42.00

No. 57 N-Nitroso Compounds: Occurrence, Biological Effects and Relevance to Human Cancer
Edited by I.K. O'Neill, R.C. von Borstel, C.T. Miller, J. Long and H. Bartsch
1984; 1013 pages; £80.00

No. 58 Age-related Factors in Carcinogenesis
Edited by A. Likhachev, V. Anisimov and R. Montesano
1985; 288 pages; £20.00

No. 59 Monitoring Human Exposure to Carcinogenic and Mutagenic Agents
Edited by A. Berlin, M. Draper, K. Hemminki and H. Vainio
1984; 457 pages; £27.50

No. 60 Burkitt's Lymphoma: A Human Cancer Model
Edited by G. Lenoir, G. O'Conor and C.L.M. Olweny
1985; 484 pages; £29.00

No. 61 Laboratory Decontamination and Destruction of Carcinogens in Laboratory Wastes: Some Haloethers
Edited by M. Castegnaro, *et al.*
1985; 55 pages; £7.50

No. 62 Directory of On-going Research in Cancer Epidemiology 1984
Edited by C.S. Muir and G. Wagner
1984; 717 pages (*out of print*)

No. 63 Virus-associated Cancers in Africa
Edited by A.O. Williams, G.T. O'Conor, G.B. de-Thé and C.A. Johnson
1984; 773 pages; £22.00

No. 64 Laboratory Decontamination and Destruction of Carcinogens in Laboratory Wastes: Some Aromatic Amines and 4-Nitrobiphenyl
Edited by M. Castegnaro, *et al.*
1985; 84 pages; £6.95

No. 65 Interpretation of Negative Epidemiological Evidence for Carcinogenicity
Edited by N.J. Wald and R. Doll
1985; 232 pages; £20.00

No. 66 The Role of the Registry in Cancer Control
Edited by D.M. Parkin, G. Wagner and C.S. Muir
1985; 152 pages; £10.00

No. 67 Transformation Assay of Established Cell Lines: Mechanisms and Application
Edited by T. Kakunaga and H. Yamasaki
1985; 225 pages; £20.00

No. 68 Environmental Carcinogens. Selected Methods of Analysis. Volume 7. Some Volatile Halogenated Hydrocarbons
Edited by L. Fishbein and I.K. O'Neill
1985; 479 pages; £42.00

No. 69 Directory of On-going Research in Cancer Epidemiology 1985
Edited by C.S. Muir and G. Wagner
1985; 745 pages; £22.00

List of IARC Publications

No. 70 **The Role of Cyclic Nucleic Acid Adducts in Carcinogenesis and Mutagenesis**
Edited by B. Singer and H. Bartsch
1986; 467 pages; £40.00

No. 71 **Environmental Carcinogens. Selected Methods of Analysis. Volume 8: Some Metals: As, Be, Cd, Cr, Ni, Pb, Se Zn**
Edited by I.K. O'Neill and, P. Schuller and L. Fishbein
1986; 485 pages; £42.00

No. 72 **Atlas of Cancer in Scotland, 1975-1980. Incidence and Epidemiological Perspective**
Edited by I. Kemp, P. Boyle, M. Smans and C.S. Muir
1985; 285 pages; £35.00

No. 73 **Laboratory Decontamination and Destruction of Carcinogens in Laboratory Wastes: Some Antineoplastic Agents**
Edited by M. Castegnaro, *et al.*
1985; 163 pages; £10.00

No. 74 **Tobacco: A Major International Health Hazard**
Edited by D. Zaridze and R. Peto
1986; 324 pages; £20.00

No. 75 **Cancer Occurrence in Developing Countries**
Edited by D.M. Parkin
1986; 339 pages; £20.00

No. 76 **Screening for Cancer of the Uterine Cervix**
Edited by M. Hakama, A.B. Miller and N.E. Day
1986; 315 pages; £25.00

No. 77 **Hexachlorobenzene: Proceedings of an International Symposium**
Edited by C.R. Morris and J.R.P. Cabral
1986; 668 pages; £50.00

No. 78 **Carcinogenicity of Alkylating Cytostatic Drugs**
Edited by D. Schmähl and J.M. Kaldor
1986; 337 pages; £25.00

No. 79 **Statistical Methods in Cancer Research. Volume III: The Design and Analsis of Long-term Animal Experiments**
Edited by J.J. Gart, D. Krewski, P.N. Lee, R.E. Tarone and J. Wahrendorf
1986; 213 pages; £20.00

No. 80 **Directory of On-going Research in Cancer Epidemiology 1986**
Edited by C.S. Muir and G. Wagner
1986; 805 pages; £22.00

No. 81 **Environmental Carcinogens: Methods of Analysis and Exposure Measurement. Volume 9: Passive Smoking**
Edited by I.K. O'Neill, K.D. Brunnemann, B. Dodet and D. Hoffmann
1987; 383 pages; £35.00

No. 82 **Statistical Methods in Cancer Research. Volume II: The Design and Analysis of Cohort Studies**
By N.E. Breslow and N.E. Day
1987; 404 pages; £30.00

No. 83 **Long-term and Short-term Assays for Carcinogens: A Critical Appraisal**
Edited by R. Montesano, H. Bartsch, H. Vainio, J. Wilbourn and H. Yamasaki
1986; 575 pages; £48.00

No. 84 **The Relevance of N-Nitroso Compounds to Human Cancer: Exposure and Mechanisms**
Edited by H. Bartsch, I.K. O'Neill and R. Schulte-Hermann
1987; 671 pages; £50.00

No. 85 **Environmental Carcinogens: Methods of Analysis and Exposure Measurement. Volume 10: Benzene and Alklated Benzenes**
Edited by L. Fishbein and I.K. O'Neill
1988; 327 pages; £35.00

No. 86 **Directory of On-going Research in Cancer Epidemiology 1987**
Edited by D.M. Parkin and J. Wahrendorf
1987; 676 pages; £22.00

No. 87 **International Incidence of Childhood Cancer**
Edited by D.M. Parkin, C.A. Stiller, C.A. Bieber, G.J. Draper. B. Terracini and J.L. Young
1988; 401 pages; £35.00

No. 88 **Cancer Incidence in Five Continents Volume V**
Edited by C. Muir, J. Waterhouse, T. Mack, J. Powell and S. Whelan
1987; 1004 pages; £50.00

No. 89 **Method for Detecting DNA Damaging Agents in Humans: Applications in Cancer Epidemiology and Prevention**
Edited by H. Bartsch, K. Hemminki and I.K. O'Neill
1988; 518 pages; £45.00

No. 90 **Non-occupational Exposure to Mineral Fibres**
Edited by J. Bignon, J. Peto and R. Saracci
1989; 500 pages; £45.00

No. 91 **Trends in Cancer Incidence in Singapore 1968-1982**
Edited by H.P. Lee, N.E. Day and K. Shanmugaratnam
1988; 160 pages; £25.00

No. 92 **Cell Differentiation, Genes and Cancer**
Edited by T. Kakunaga, T. Sugimura, L. Tomatis and H. Yamasaki
1988; 204 pages; £25.00

List of IARC Publications

No. 93 Directory of On-going Research in Cancer Epidemiology 1988
Edited by M. Coleman and J. Wahrendorf
1988; 662 pages (*out of print*)

No. 94 Human Papillomavirus and Cervical Cancer
Edited by N. Muñoz, F.X. Bosch and O.M. Jensen
1989; 154 pages; £19.00

No. 95 Cancer Registration: Principles and Methods
Edited by O.M. Jensen, D.M. Parkin, R. MacLennan, C.S. Muir and R. Skeet
Publ. due 1990; approx. 300 pages

No. 96 Perinatal and Multigeneration Carcinogenesis
Edited by N.P. Napalkov, J.M. Rice, L. Tomatis and H. Yamasaki
1989; 436 pages; £48.00

No. 97 Occupational Exposure to Silica and Cancer Risk
Edited by L. Simonato, A.C. Fletcher, R. Saracci and T. Thomas
Publ. due 1990; approx. 160 pages; £19.00

No. 98 Cancer Incidence in Jewish Migrants to Israel, 1961-1981
Edited by R. Steinitz, D.M. Parkin, J.L. Young, C.A. Bieber and L. Katz
1988; 320 pages; £30.00

No. 99 Pathology of Tumours in Laboratory Animals, Second Edition, Volume 1, Tumours of the Rat
Edited by V.S. Turusov and U. Mohr
Publ. due 1990; approx. 700 pages; £85.00

No. 100 Cancer: Causes, Occurrence and Control
Edited by L. Tomatis
1990; approx. 350 pages; £24.00

No. 101 Directory of On-going Research in Cancer Epidemiology 1989-90
Edited by M. Coleman and J. Wahrendorf
1989; 818 pages; £36.00

No. 102 Patterns of Cancer in Five Continents
Edited by S.L. Whelan and D.M. Parkin
Publ. due 1990; approx. 150 pages; £25.00

No. 103 Evaluating Effectiveness of Primary Prevention of Cancer
Edited by M. Hakama, V. Beral, J.W. Cullen and D.M. Parkin
Publ. due 1990; approx. 250 pages; £32.00

No. 104 Complex Mixtures and Cancer Risk
Edited by H. Vainio, M. Sorsa and A.J. McMichael
Publ. due 1990; approx. 450 pages; £38.00

No. 105 Relevance to Human Cancer of N-Nitroso Compounds, Tobacco Smoke and Mycotoxins
Edited by I.K. O'Neill, J. Chen, S.H. Lu and H. Bartsch
Publ. due 1990; approx. 600 pages

IARC MONOGRAPHS ON THE EVALUATION OF CARCINOGENIC RISKS TO HUMANS

(Available from booksellers through the network of WHO Sales Agents*)

Volume 1 **Some Inorganic Substances, Chlorinated Hydrocarbons, Aromatic Amines, N-nitroso Compounds, and Natural Products**
1972; 184 pages (*out of print*)

Volume 2 **Some Inorganic and Organometallic Compounds**
1973; 181 pages (*out of print*)

Volume 3 **Certain Polycyclic Aromatic Hydrocarbons and Heterocyclic Compounds**
1973; 271 pages (*out of print*)

Volume 4 **Some Aromatic Amines, Hydrazine and Related Substances, N-nitroso Compounds and Miscellaneous Alkylating Agents**
1974; 286 pages;
Sw. fr. 18.–/US $14.40

Volume 5 **Some Organochlorine Pesticides**
1974; 241 pages (*out of print*)

Volume 6 **Sex Hormones**
1974; 243 pages (*out of print*)

Volume 7 **Some Anti-Thyroid and Related Substances, Nitrofurans and Industrial Chemicals**
1974; 326 pages (*out of print*)

Volume 8 **Some Aromatic Azo Compounds**
1975; 375 pages;
Sw. fr. 36.–/US $28.80

Volume 9 **Some Aziridines, N-, S- and O-Mustards and Selenium**
1975; 268 pages;
Sw. fr. 27.–/US $21.60

Volume 10 **Some Naturally Occurring Substances**
1976; 353 pages (*out of print*)

Volume 11 **Cadmium, Nickel, Some Epoxides, Miscellaneous Industrial Chemicals and General Considerations on Volatile Anaesthetics**
1976; 306 pages (*out of print*)

Volume 12 **Some Carbamates, Thiocarbamates and Carbazides**
1976; 282 pages;
Sw. fr. 34.–/US $27.20

Volume 13 **Some Miscellaneous Pharmaceutical Substances**
1977; 255 pages;
Sw. fr. 30.–/US $24.00

Volume 14 **Asbestos**
1977; 106 pages (*out of print*)

Volume 15 **Some Fumigants, The Herbicides 2,4-D and 2,4,5-T, Chlorinated Dibenzodioxins and Miscellaneous Industrial Chemicals**
1977; 354 pages;
Sw. fr. 50.–/US $40.00

Volume 16 **Some Aromatic Amines and Related Nitro Compounds – Hair Dyes, Colouring Agents and Miscellaneous Industrial Chemicals**
1978; 400 pages;
Sw. fr. 50.–/US $40.00

Volume 17 **Some N-Nitroso Compounds**
1987; 365 pages;
Sw. fr. 50.–/US $40.00

Volume 18 **Polychlorinated Biphenyls and Polybrominated Biphenyls**
1978; 140 pages;
Sw. fr. 20.–/US $16.00

Volume 19 **Some Monomers, Plastics and Synthetic Elastomers, and Acrolein**
1979; 513 pages;
Sw. fr. 60.–/US $48.00

Volume 20 **Some Halogenated Hydrocarbons**
1979; 609 pages (*out of print*)

Volume 21 **Sex Hormones (II)**
1979; 583 pages;
Sw. fr. 60.–/US $48.00

Volume 22 **Some Non-Nutritive Sweetening Agents**
1980; 208 pages;
Sw. fr. 25.–/US $20.00

Volume 23 **Some Metals and Metallic Compounds**
1980; 438 pages (*out of print*)

Volume 24 **Some Pharmaceutical Drugs**
1980; 337 pages;
Sw. fr. 40.–/US $32.00

Volume 25 **Wood, Leather and Some Associated Industries**
1981; 412 pages;
Sw. fr. 60.–/US $48.00

Volume 26 **Some Antineoplastic and Immunosuppressive Agents**
1981; 411 pages;
Sw. fr. 62.–/US $49.60

Volume 27 **Some Aromatic Amines, Anthraquinones and Nitroso Compounds, and Inorganic Fluorides Used in Drinking Water and Dental Preparations**
1982; 341 pages;
Sw. fr. 40.–/US $32.00

Volume 28 **The Rubber Industry**
1982; 486 pages;
Sw. fr. 70.–/US $56.00

---

* A list of WHO sales agents may be obtained from the Distribution of Sales Service, World Health Organization, 1211 Geneva 27, Switzerland

List of IARC Publications

Volume 29 **Some Industrial Chemicals and Dyestuffs**
1982; 416 pages;
Sw. fr. 60.-/US $48.00

Volume 30 **Miscellaneous Pesticides**
1983; 424 pages;
Sw. fr. 60.-/US $48.00

Volue 31 **Some Food Additives, Feed Additives and Naturally Occurring Substances**
1983; 314 pages;
Sw. fr. 60.-/US $48.00

Volume 32 **Polynuclear Aromatic Compounds, Part 1: Chemical, Environmental and Experimental Data**
1984; 477 pages;
Sw. fr. 60.-/US $48.00

Volume 33 **Polynuclear Aromatic Compounds, Part 2: Carbon Blacks, Mineral Oils and Some Nitroarenes**
1984; 245 pages;
Sw. fr. 50.-/US $40.00

Volume 34 **Polynuclear Aromatic Compounds, Part 3: Industrial Exposures in Aluminium Production, Coal Gasification, Coke Production, and Iron and Steel Founding**
1984; 219 pages;
Sw. fr. 48.-/US $38.40

Volume 35 **Polynuclear Aromatic Compounds: Part 4: Bitumens, Coal-Tars and Derived Products, Shale-Oils and Soots**
1985; 271 pages;
Sw. fr. 70.-/US $56.00

Volume 36 **Allyl Compounds, Aldehydes, Epoxides and Peroxides**
1985; 369 pages;
Sw. fr. 70.-/US $56.00

Volume 37 **Tobacco Habits Other than Smoking; Betel-Quid and Areca-Nut Chewing; and Some Related Nitrosamines**
1985; 291 pages;
Sw. fr. 70.-/US $56.00

Volume 38 **Tobacco Smoking**
1986; 421 pages;
Sw. fr. 75.-/US $60.00

Volume 39 **Some Chemicals Used in Plastics and Elastomers**
1986; 403 pages;
Sw. fr. 60.-/US $48.00

Volume 40 **Some Naturally Occurring and Synthetic Food Components, Furocoumarins and Ultraviolet Radiation**
1986; 444 pages;
Sw. fr. 65.-/US $52.00

Volume 41 **Some Halogenated Hydrocarbons and Pesticide Exposures**
1986; 434 pages;
Sw. fr. 65.-/US $52.00

Volume 42 **Silica and Some Silicates**
1987; 289 pages;
Sw. fr. 65.-/US $52.00

Volume 43 **Man-Made Mineral Fibres and Radon**
1988; 300 pages;
Sw. fr. 65.-/US $52.00

Volume 44 **Alcohol Drinking**
1988; 416 pages;
Sw. fr. 65.-/US $52.00

Volume 45 **Occupational Exposures in Petroleum Refining; Crude Oil and Major Petroleum Fuels**
1989; 322 pages;
Sw. fr. 65.-/US $52.00

Volume 46 **Diesel and Gasoline Engine Exhausts and Some Nitroarenes**
1989; 458 pages;
Sw. fr. 65.-/US $52.00

Volume 47 **Some Organic Solvents, Resin Monomers and Related Compounds, Pigments and Occupational Exposures in Paint Manufacture and Painting**
1990; 536 pages;
Sw. fr. 85.-/US $ 68.00

Volume 48 **Some Flame Retardants and Textile Chemicals, and Exposures in the Textile Manufacturing Industry**
1990; approx. 350 pages;
Sw. fr. 65.-/US $52.00

Supplement No. 1
Chemicals and industrial processes associated with cancer in humans (IARC Monographs, Volumes 1 to 20)
1979; 71 pages; (*out of print*)

Supplement No. 2
Long-Term and Short-Term Screening Assays for Carcinogens: A critical Appraisal
1980; 426 pages;
Sw. fr. 40.-/US $32.00

Supplement No. 3
Cross index of synonyms and trade names in Volumes 1 to 26
1982; 199 pages (*out of print*)

Supplement No. 4
Chemicals, industrial processes and industries associated with cancer in humans (IARC Monographs, Volumes 1 to 29)
1982; 292 pages (*out of print*)

Supplement No. 5
Cross Index of Synonyms and Trade Names in Volumes 1 to 36
1985; 259 pages;
Sw. fr. 46.-/US $36.80

Supplement No. 6
Genetic and Related Effects: An Updating of Selected IARC Monographs from Volumes 1 to 42
1987; 729 pages;
Sw. fr. 80.-/US $64.00

List of IARC Publications

Supplement No. 7
Overall Evaluations of Carcinogenicity: An Updating of IARC Monographs Volumes 1–42
1987; 434 pages; Sw. fr. 65.–/US $52.00

Supplement No. 8
Cross Index of Synonyms and Trade Names in Volumes 1 to 46 of the IARC Monographs
Publ. 1990; 260 pages; Sw. fr. 60.–/US $48.00

IARC TECHNICAL REPORTS*

No. 1 **Cancer in Costa Rica**
Edited by R. Sierra, R. Barrantes, G. Muñoz Leiva, D.M. Parkin, C.A. Bieber and N. Muñoz Calero
1988; 124 pages; Sw. fr. 30.–/US $24.00

No. 2 **SEARCH: A Computer Package to Assist the Statistical Analysis of Case–Control Studies**
Edited by G.J. Macfarlane, P. Boyle and P. Maisonneuve (in press)

No. 3 **Cancer Registration in the European Economic Community**
Edited by M.P. Coleman and E. Démaret
1988; 188 pages; Sw. fr. 30.–/US $24.00

No. 4 **Diet, Hormones and Cancer: Methodological Issues for Prospective Studies**
Edited by E. Riboli and R. Saracci
1988; 156 pages; Sw. fr. 30.–/US $24.00

No. 5 **Cancer in the Philippines**
Edited by A.V. Laudico, D. Esteban and D.M. Parkin
1989; 186 pages; Sw. fr.30.–/US $24.00

INFORMATION BULLETINS ON THE SURVEY OF CHEMICALS BEING TESTED FOR CARCINOGENICITY*

No. 8 **Edited by M.-J. Ghess, H. Bartsch and L. Tomatis**
1979; 604 pages; Sw. fr. 40.–

No. 9 **Edited by M.-J. Ghess, J.D. Wilbourn, H. Bartsch and L. Tomatis** 1981; 294 pages; Sw. fr. 41.–

No. 10 **Edited by M.-J. Ghess, J.D. Wilbourn and H. Bartsch** 1982; 362 pages; Sw. fr. 42.–

No. 11 **Edited by M.-J. Ghess, J.D. Wilbourn, H. Vainio and H. Bartsch**
1984; 362 pages; Sw. fr. 50.–

No. 12 **Edited by M.-J. Ghess, J.D. Wilbourn, A. Tossavainen and H. Vainio**
1986; 385 pages; Sw. fr. 50.–

No. 13 **Edited by M.-J. Ghess, J.D. Wilbourn and A. Aitio** 1988; 404 pages; Sw. fr. 43.

NON–SERIAL PUBLICATIONS †

Alcool et Cancer
By A. Tuyns (in French only)
1978; 42 pages; Fr. fr. 35.–

Cancer Morbidity and Causes of Death Among Danish Brewery Worker
By O.M. Jensen 1980; 143 pages; Fr. fr. 75.–

Directory of Computer Systems Used in Cancer Registries By H.R. Menck and D.M. Parkin
1986; 236 pages; Fr. fr. 50.–

*Available from booksellers through the network of WHO sales agents.

† Available directly from IARC.